Feeling Climate Change

Examining the social response to the mounting impacts of climate change, *Feeling Climate Change* illuminates what the pathways from emotions to social change look like—and how they work—so we can recognize and inform our collective attempts to avert further climate catastrophe.

Debra J. Davidson engages with how our actions are governed by a complex of rules, norms, and predispositions, central among which operates our emotionality, to assess individual and collective responses to the climate crisis, applying a critical and constructive analysis of human social prospects for confronting the climate emergency in manners that minimize the damage and perhaps even enhance the prospects for meaningful collective living.

Providing a crucial understanding of our emotionality and its role in individual behaviour, collective action, and ultimately in social change, this book offers researchers, policymakers, and citizens essential insights into our personal and collective responses to the climate emergency.

Debra J. Davidson is professor of environmental sociology at the University of Alberta. She is co-editor of the *Oxford Handbook of Energy and Society* (2018) and co-editor of *Environment and Society* (2018), as well as author of numerous articles on sociology and the environment.

Feeling Climate Change
How Emotions Govern Our Responses to the Climate Emergency

Debra J. Davidson

Routledge
Taylor & Francis Group

NEW YORK AND LONDON

Designed cover image: 135 fires (original is linocut, screen print on paper; 7x5"), by Lori Claerhout

First published 2025
by Routledge
605 Third Avenue, New York, NY 10158

and by Routledge
4 Park Square, Milton Park, Abingdon, Oxon, OX14 4RN

Routledge is an imprint of the Taylor & Francis Group, an informa business

© 2025 Debra J. Davidson

The right of Debra J. Davidson to be identified as author of this work has been asserted in accordance with sections 77 and 78 of the Copyright, Designs and Patents Act 1988.

All rights reserved. No part of this book may be reprinted or reproduced or utilised in any form or by any electronic, mechanical, or other means, now known or hereafter invented, including photocopying and recording, or in any information storage or retrieval system, without permission in writing from the publishers.

Trademark notice: Product or corporate names may be trademarks or registered trademarks, and are used only for identification and explanation without intent to infringe.

Library of Congress Cataloging-in-Publication Data
Names: Davidson, Debra J., author.
Title: Feeling climate change: how emotions govern our responses to the climate emergency / Debra J. Davidson.
Description: New York, NY: Routledge, 2024. | Includes bibliographical references and index.
Identifiers: LCCN 2024015627 | ISBN 9781032462813 (hardback) | ISBN 9781032462769 (paperback) | ISBN 9781003380900 (ebook)
Subjects: LCSH: Climatic changes–Psychological aspects.
Classification: LCC BF353.5.C55 D385 2024 | DDC 155.9/15–dc23/eng/20240508
LC record available at https://lccn.loc.gov/2024015627

ISBN: 978-1-032-46281-3 (hbk)
ISBN: 978-1-032-46276-9 (pbk)
ISBN: 978-1-003-38090-0 (ebk)

DOI: 10.4324/9781003380900

Typeset in Sabon
by Deanta Global Publishing Services, Chennai, India

Contents

	Acknowledgements	*viii*
1	Introduction: why a book on emotions?	1
2	What Lies Ahead	12
3	Can We Do This? Embarking on Transformational Social Change	38
4	What Are Emotions and Why Should We Care?	57
5	Scaling Up Emotions, from the Individual, to Social Structures and Back Again	82
6	Inaction Pathways: On Why We Don't Do the Things We Don't Do	104
7	Pathways to Action, or Doing the Hard Thing	129
8	Threading the Needle from Emotions to Transformational Social Change	154
	Works Cited	*167*
	Index	*192*

Acknowledgements

Collecting and sharing knowledge is never an individual affair, and I have many people to thank for helping me through this process, from lofty idea to books in print. First, I wish to thank my research collaborators, Maik Kecinski at the University of Delaware and Kyle Nash at the University of Alberta, for their commitments to our ongoing research focused on emotional responses to climate change. I've also received generous encouragement and support from several amazing friends and colleagues, including Amy Kaler, Mike Gismondi, Elizabeth Ho, Gwendolyn Blue, Christina Hoicka, Carrie Karsgaard, and Lianne Lefsrud. I am so blessed to have each of you in my circle!

A very big thanks as well to my graduate students, who always ask the right questions, and share unique perspectives, including in particular Angeline Letourneau, Kaan Ozdurak, Joseph Flesch, Jessica Hermary, and Angelina Fedorenko. I have also received tremendous support from my home department, the Department of Resource Economics and Environmental Sociology at the University of Alberta, not least for granting me a year of sabbatical during which much of this book was written. And I wish to thank Michael Gibson, Senior Sociology Editor with Routledge, for his generous encouragement and assistance.

I want to extend a special thanks as well to my friend, fellow hiker and amazing artist extraordinaire Lori-Ann Claerhout, for her willingness to explore an artist-academic collaboration in this book, to see if we can encourage our readers to open their artistic as well as intellectual ways of knowing as they make their way through the coming chapters. Here is how Lori describes her approach: 'The through line is around knowing that we need each and all of us to move through our emotional states and everything else around the climate crisis, so I'm looking to indicate "collectivity" and 'movement' with it.'

Finally, I wish to thank my daughters, Miko and Kamoura, for their patience, support, occasionally brutal honesty, and love.

Cover Image: 135 fires (original is linocut, screen print on paper; 7x5"), by Lori Claerhout.

On the day of creating this print, there were 135 fires actively burning across Alberta. The province of Alberta recorded over a thousand fires in 2023's fire season. Over two million hectares burned there of a total 18.5 million hectares burned across Canada.

Lori Claerhout also provided the illustrations included at the beginning of each chapter. I asked Lori to first read the chapter, and then develop an image that captures her interpretation of the chapter's main themes. These images consist of open line work and absence of shading or colour work which considers body text while suggesting themes and allowing space for imagining alternatives of collectivity and movement. I love the result, and I hope you do too!

Lori Claerhout is a print and found objects artist based in northern Alberta, Canada. Broadly, her work is concerned with social and environmental issues and seeing what's familiar in new ways; to visually critique often overlooked issues that might also contain significance to our existence.

You can learn more about Lori and her work at www.loriclaerhout.com

1 Introduction
why a book on emotions?

"Passion is our essential power vigorously striving to attain its object."
Karl Marx, Economic and Philosophical Manuscripts, 1844.

This book began as a disorganized file box of mental notes and personal musings that emerged from my own engagement with the climate emergency, and even more so, my engagement with others in discussions about the climate emergency. These musings, juxtaposed against my academic knowledge of global warming and human behaviour, suggested some rather deep disconnects. That passion Marx speaks of, referring to our emotionality, is an open secret in academic and policy circles: we all know experientially that we humans are highly emotional beings, and that these emotions govern our actions. And yet, scientists and policymakers alike tend to approach our personal and collective responses to the climate emergency, as we are trained to

DOI: 10.4324/9781003380900-1

do, with a strictly cerebral lens—we researchers must set our own emotions aside, and we presume that everyone else does too. Responses to information about global warming are defined by knowledge levels, instrumental resources, self-interest, and our demographic profiles, we are told. The reason I am so alarmed about the climate emergency, thus, is because I am a well-informed, white, middle class, bleeding heart liberal mother of two who would like to be able to enjoy a pristine environment. But, like you, I know many such people who share the exact same profile as me, who nonetheless respond to the climate emergency, and other goings on in our lives, in entirely different ways. We are not, after all, simply automatons activated by instructions provided by our socio-structural context. We each absorb much from these instructions to be sure, but we also interpret, navigate, rail against, and influence those social structures, on the basis of our personal life experiences, identities and commitments.

There was another observation that struck me even more bluntly—my reactions to global warming information appeared to differ from others most starkly on an emotional level. This became most vivid when Greta came to town. Yes, Great made a stopover in Edmonton, Alberta while on her North American tour! I commandeered my kids into helping me paint slogans onto cardboard and glue them to wooden posts, and we hopped on the train and converged with several thousand other Albertans on the Legislature grounds to show our support for Greta Thunberg, then-15-year-old leader of the Fridays for Future movement, who has inspired millions of young people around the globe to stand up to those elite forces determined to keep us on a trajectory of skyrocketing global average temperatures. There was a lot of excitement and hope on display that sunny autumn day; Greta called more Albertans from their homes to the Legislature than had shown up for a climate protest, or any other protest for that matter, in years, perhaps decades (Albertans are not known for their street protests).

But that was not all. Snaking through the crowds of the giddy and the merely curious were a significant—significantly loud anyway—minority. These attendees were not anticipating Greta's arrival with excitement, in fact they weren't even looking up at the top of the staircase where she would soon emerge. Instead, they were yelling, angrily, at me. And others. They were yelling about how Greta should go back home. Did I know her coat is made of petrochemical products? Didn't I know that Alberta's oil is the best, cleanest, most ethical oil in the world? We sheeple have been duped by the climate hoax and we need to wake up. Around the perimeter of the Legislative grounds, meanwhile, the sounds of horns from semi-trucks were blaring—a harbinger of the right-wing Trucker Convoy that would bring the nation's capital to the brink of breakdown a short few years later—some having traveled hours in their flag-dressed shiny big rigs up Highway 2 to attempt to disrupt the speeches that day, to put a stop to all this offensive, oil-worker-hating, climate action nonsense.

This soft- but clearly-spoken young woman, who was in essence sharing what was by then the consensus scientific view on climate change, managed to motivate a level of enthusiasm and excitement among the climate concerned—even in quiet, industrious and pragmatic Edmonton—that was desperately lacking in the climate movement until then, and at the same time generate bright red, spit-slewing rage in others, bringing fellow Albertans very nearly to blows. I could no longer simply ponder this disconnect between those academic treatments of climate beliefs and behaviour worked up in our offices and laboratories, and the climate behaviour that was erupting around me, in the streets and on our smart phones. Because, after all, if our collective efforts to motivate progressive action to address the climate emergency are ultimately missing some of the most crucial elements that influence our personal responses to this crisis, then how effective could those strategies possibly be?

So, I began to read, and read, and read some more, and found myself in a rabbit hole from which I have yet to emerge. I read everything sociologists have to say about emotions, and then I read psychology, neuroscience, human evolution, and media studies. When read together, the message was clear: we in the climate science and advocacy communities are missing something, something important, that can enhance our understanding of just how deep is the human and social toll of global warming, but even more importantly, some essential ingredients for our collective confrontation with it. My time in that rabbit hole has led to new research projects, ongoing as of this writing, and many other researchers have begun to open the black box of climate emotionality as well. The findings that have emerged provide some valuable insights, and raise yet more questions.

That most vexing of questions at this point in time is not *Is climate change real?* Or even *How will climate change affect us?* Rather, it is *Why have we not responded in a manner requisite with the scale of the emergency?* Numerous structural barriers have shared the explanatory limelight: the lack of political will cultivated in short-term electoral cycles; the power of the corporate elite; the 'lock-in' effects of fossil fuel-based energy infrastructure. These answers are not wrong, but they aren't fully satisfactory either. After all, there are plenty of regions around the globe where one or more of these is absent, yet inaction still prevails, and others in which we observe the opposite. These structural arguments, I argue, aren't getting to the heart of the matter. No matter how complex the social system, the main driver of change is social actors: emotional, intelligent reflexive human beings exercising their agency. Before you roll your eyes, no this isn't a treatise on Thatcheresque individualism. The vehicle for social change has always been and will continue to be organized, *collective* action, but that collective action begins with individuals deciding to act collectively. This is even true for many seemingly autonomous levers of social change, like the internet. Many of the dramatic changes in our lives introduced by the internet were, to be sure, emergent and unanticipated. Yet the internet would not have caused such revolutionary change without extensive organized action to develop the technology and infrastructure that enabled its accessibility, and its subsequent adoption by so many users. The rise of a merchant class, similarly, was not merely an autonomic shift in economic structures, it was the result of individuals choosing to shed pre-existing occupational roles, form new associations, and engage in new trade relations. And within the academy, scientific paradigm shifts do not surface of their own accord; they are premised on researchers risking their careers to acknowledge the limitations of an incumbent paradigm, and creating and ultimately embracing a new one.

In these examples, the actors involved are not necessarily seeking social structural change; they are more often than not simply pursuing personal and collective projects with far less grandiose end goals, but their actions nonetheless have culminated in substantial change at the structural level. These emergent, seemingly unplanned forms of change in our social systems occur far more often than their counterparts: deliberate, purposeful social change toward a desired end goal, which is what is called for to respond to the climate emergency, but the mechanics are the same. All social change is premised upon organized collective action, and that collective action is initiated and sustained by *human agency*: committed individuals, reflecting upon their situation in the context of their values, experiences, and social feedback, and choosing to do things differently than before. Even the outcomes of externally-imposed disruptions—say military conquest or climatic change—are shaped in very large part by the character of agentic responses to them. Agency, nonetheless, does not operate in a vacuum. Reflexive individuals confront cultural, political and economic structures that do not simply delimit the actions of agents; they shape how we think and feel. Today, those structures defining Western social systems—notably but not exclusively capitalism, colonialism and patriarchy—reward a privileged minority while harming many, and the capacity to confront those structures is differentially distributed throughout society.

So, what are the motivations and enablements that encourage someone to get involved in, or resist, efforts to address the climate emergency, and why? Most efforts to respond to this and related questions, and there have been many such efforts, tend to embrace some form of rational-utilitarian model of human behaviour, despite its well-recognized flaws. This is beginning to change, however. Environmental social scientists have begun to embrace what neuroscientists have understood for decades: emotions are the starting point for decision-making. Human agency is rooted in emotionality. Hardly any more seen as the poor and embarrassing sibling of cognitive function, in research particularly since the turn of the 21st Century, emotions have been given their due. While we are starting to make up ground, however, with growing research interest in the role of emotions in decision-making and behaviour, there remains a tremendous need to integrate and apply what are at the moment numerous disparate theoretical and empirical threads, transcending an equally large number of disciplines, which still tend to be far too compartmentalized to allow for a comprehensive treatment of the role of emotions in social change processes.

A full understanding of emotionality and its role in social change thus demands that we embrace interdisciplinarity to support a multi-dimensional gaze: inward, to personal emotional experiences; outward, to how emotions operate in highly diverse and dynamic social systems; backward, to the origins of our emotionality; and forward, to forecast plausible future pathways given our current social and ecological context. We first need to delve into human bodies, how emotions are generated and processed neurologically, allowing us to get deeper into the minds of those individual social actors and take a look at how emotions govern their decisions and behaviours. We now understand that affective stimuli in our brains enter consciousness and govern behaviour, including social behaviour, and those behaviours in turn intervene through personal and collective action in our social structures. At the same time, we need to attend to the social context of emotionality: how our emotions are shaped by our interactions within groups, our institutions and our social structures, and those groups, institutions and structures influence our emotional responses to the world around us, in adaptive, and—often—maladaptive ways. We also need to trace current emotional states back through time to grasp their origins, and how they govern decisions today that inform future trajectories.

In doing so, we can trace threads, or causal mechanisms, directly from socio-ecological context, to human emotions, to social change and back again. Indeed, the extraordinary success of the human species in enduring all manner of ecological disruption, and adapting to nearly every habitat on earth, is due not to our superior tool-making skills, not to our technological expertise, but rather to our ability to cooperate, to learn, *from each other*. The conditions allowing for shared learning, and for purposeful social change, are brought forth by our emotionality. Our emotionality has been the cause of all manner of calamity, but it is also our route to collective problem solving.

Of course, nothing is simple. Understanding emotions does not give us the power to predict human behaviour, because the relationships between emotions and behaviour are by no means deterministic. Our complex emotional palette is inherited, the outcome of interactions between nature and culture throughout human history. And yet, although the genetic make-up of the human species is strikingly consistent, because of the interaction of inherited traits with local social and environmental cultures, with a bit of genetic variation thrown in, we are not all in fact the same. While on the one hand, for example, we all have the ability to feel embarrassment or shame, the interpretation of circumstances as shameful will vary substantially, based on culture and intersectional positionality. We all also express differences in emotional inclinations, called affective traits. We all know this to be true experientially—some people are just more likely to be happy, to be angry, or more or less prone to show empathy toward others. And finally, we are each conscious, reflexive agents, perfectly capable of following scripts—whether those scripts were written by our hind brains or by our cultural context—but we are also, *thankfully,* capable of deviating from them.

The main thesis of book is that we have the capacity to navigate the climate emergency in a manner that, while we won't be unscathed, transforms our current, highly toxic relationship with our climate into one that is more mutually respectful. Future pathways are not yet foretold; those routes will be mapped by actions taken now, individually, and even more so collectively. Historic emissions have already unleashed shifts throughout our biosphere the import of which we are only beginning to understand. Meanwhile, resistance to aggressive mitigation strategies orchestrated by the beneficiaries of the current, fossil-fueled neoliberal capitalist global economy remains formidable, promising further insult to injury as the political struggles to respond to the climate emergency endure. Neither of these realities call for throwing in the towel, however. To the contrary, they are calls to action. Maximizing our capacities to get through this moment through collective engagement, however, must proceed not by ignoring or rejecting our emotionality but by working with it.

There are a few important points of clarification, and qualification, before we proceed. First, a note on terms used in this book. I use the term *climate change* to refer to the subject of science; the findings generated from this body of research have indicated that we are currently experiencing *global warming*; and given what we now understand to be the implications of that warming for our planetary systems and all its inhabitants, we are facing a *climate emergency.*

Second, who is *we?* In many places throughout this book, I use the term 'we' to refer to all of us, the human species. I do so to highlight two things: our commonalities, particularly when it comes to emotional and cognitive processes; and the threat that global warming poses to all of humanity. However, I do so with the full understanding that underlying this 'we' are multiple layers of difference. A critical, intersectional lens is essential to allow for deeper understanding of the climate emergency—how we got here and where we are going—but also in order to reflect upon variations in emotionality according to age, gender, class, culture, ethnicity, Indigeneity and sexual identity. Most notably, today's social structures impose enormous differences in accountability for causing the climate emergency, differences in vulnerability to global warming's effects, and differences in capability to adapt to those effects, as well as to engage in personal and collective action to address the climate emergency. These differences matter, and recognizing differences must serve as the foundation for a politics of change. This also means not everyone is equally responsible for doing the work that is required. Having the power to influence is a source of privilege. It is also a responsibility. Therefore, in my calls to action, I am not invoking the universal We; I am looking specifically to those groups that are to differing degrees 'part of the problem,' and/or are in a position to be 'part of the solution.' At the same time, I will argue that one of the most important things each of us—but particularly those groups just mentioned—can do in order to lay the groundwork for transformational social change is to confront those differences, to shift the boundaries of our in-groups to include more people who are not just like us, to invoke the We that is our common humanity.

Third and finally, this book represents *my* truth, my understanding of the climate emergency and how to confront it, but there are many others. I am a middle-aged, white, queer mom and life-long environmentalist, born into a lower middle class, two household family in North America, and the only one of three siblings who expressed a desire to go to university, where I was trained first in environmental studies and then in sociology. My unique set of life experiences indelibly shapes my perspective, the things I understand, and the things I don't. No single version of reality should be received as absolute, and I encourage readers to seek out as many different perspectives as possible. Academics do not have the penultimate insight into our complex world, and neither do white folks, or North Americans. Unfortunately, we white, North American academics nonetheless take up a lot of the air in climate dialogues, and I particularly encourage readers to look beyond these voices to hear other valuable truths.

Outline of the Chapters to Follow

This book is divided into eight chapters, covering the following questions: Just how serious is the mess we have created, and are we capable of making the changes needed to prevent things from getting even worse, and maybe even making some things better? What are emotions and how do they feature in those change-making efforts? How can an emotionality lens help to elucidate observed social responses to climate change? Namely, what differentiates emotional pathways toward inaction and toward action? Most importantly, how can we apply this knowledge to facilitate the transformational changes required to support social and ecological wellbeing in our new climate reality?

I begin this conversation in Chapter 2 with a brief snapshot of current climate science, at the previously inconceivable disruptions that have already begun to materialize, and some pretty scary future projections. This informs a clear-eyed acknowledgement that the path forward will be fraught with economic disruptions, escalating disasters, political upheaval, and unbearable loss, all of which will manifest in starkly inequitable ways. Engaging collectively at the scale required to confront the climate emergency, while simultaneously coming to terms with substantial losses of lives, livelihoods and things of value will be formidable. What are the implications for civil society of enduring anxiety-, loss- and trauma-filled days, months, years, at the same time at which substantial personal investments in collective action are called for? With so many opportunities for fatalism and escape available to us, where lies the motivation and capacity to confront the dragon: a fossil-fueled neoliberal capitalist economic system and its wealthy beneficiaries?

With the scale of the challenge laid out before us, Chapter 3 then asks, *Can we do this?* Is it even realistic to expect the transformational changes required to be realized? The answer I offer in this chapter, on the basis of reviewing decades of research on social change from various corners of the academy, is *maybe*, which is not exactly comforting, but it is enough to make trying worthwhile, given the alternative. While it is safe to say that the scale of structural change that has already begun to unfold is likely unprecedented in human history, the possible organizational responses to those destabilizing forces nonetheless premise multiple potential scenarios, from orchestrated transition to low-emission socio-economic systems on one end of the spectrum, to catastrophic loss of life and social breakdown on the other. In this chapter, readers will also get a primer on structures, institutions, and agency, and my conceptual understanding of how they fit together, which also offers support for my argument that the potential for transformational change exists, and also points the finger at the key change agent: you and me.

The next two chapters bring emotions to the fore, beginning with a detailed overview of just what emotions are, why we have them, and what they do, in Chapter 4. This Chapter draws from an expansive record of theoretical and empirical works generated by scholars in the affective sciences, behavioural economics, and sociology, laying the groundwork for the remainder of the book. Topics of discussion include the critical role played by emotions in decision-making, and the evolutionary origins of our emotions: the means by which our inherited emotion-cognition repertoire represents attributes that have been selected and retained throughout human history because they have at some point had positive bearing on our survival. More recently, this includes ways that help us navigate our increasingly complex social worlds. Those predispositions are Janus-faced, however: many of those emotion-cognition faculties that have served us so well in the past can support things like xenophobia, climate denial, or just resistance to change when change is very much called for. A subset of emotions will be explored in further detail, based on their import for social responses to climate change. These include pride, shame, guilt, and empathy, the latter of which is crucial to cooperative responses to collective action problems.

In Chapter 5, this conversation about emotions is scaled up, from the individual-level, which served as the primary unit of analysis in Chapter 4, to consider institutional and structural contexts; contexts that provide feedstock for our emotional responses, and at the same time reward and sanction our emotionality. We manage our emotions accordingly, and respond to emerging stimuli, such as climate change information, through these lenses. Three structures are highlighted that are particularly germane—capitalism, colonialism, and patriarchy—which generate conditions like oppression, inequity, precarity, and ecological disruption, all of which in turn affect our emotionality in ways that are consequential to social responses to the climate emergency. Ultimately, however, despite our elaborate, complex social structures, we each reflect upon, and navigate these structures along emotion-cognition pathways that ultimately guide some of us toward maintaining those structures, and others toward collective commitments to challenge them.

Chapters 6 and 7 get to the heart of the matter: In these chapters, the knowledge shared in Chapters 4 and 5 will be applied to observed social responses to the climate emergency. Individual responses to the climate emergency involve threat detection, evaluation of response options, and pursuit of pro-climate action (or not). Emotions thread their way through this entire process. Some emotion-cognition pathways support collective engagement and social change, but many favour inaction. The now extensive empirical literature on this topic will be synthesized and re-examined through an emotionality lens, to demarcate emotional pathways that favor inaction and those that favor pro-climate action, focusing both on our inherited emotional predispositions, and structural conditions that facilitate certain emotional and behavioural responses over others. In Chapter 6, four pathways to inaction are discussed, including apathy, denial, withdrawal, and 'stuck'. There are far more people across the globe have *not* responded to the climate emergency—personally or politically—than have, suggesting that our emotional responses to climate change are more likely than not to support inaction. Failure to detect a stimulus at all will transpire as apathy. Global warming is effectively invisible to our senses, and therefore our 'detection' of the threat entails accessing and accepting information from scientists, the media, and opinion leaders—a route that has not been particularly well-honed in our affective repertoire. The means by which this information is conveyed, and by whom, consequently matters a whole lot. In all too many cases, scientific information about global warming has conveyed what is for many an abstract, temporally and spatially distant phenomenon, unlikely to capture attention until individuals come to recognize how it threatens the things they value. Lacking experience with impacts that are attributed to climate change, individuals tend to assign an erroneously small risk perception to such threats, particularly when our attention is otherwise occupied by busy, stressful lives, describing many people in Western countries.

10 Introduction

By contrast, denial describes a coping mechanism invoked when we detect stimuli that we find too emotionally overwhelming to process. In other words, all forms of climate denial represent an acutely emotional response. Global warming for many constitutes a threat—to beliefs systems, to social identities, and to lifestyles—so overwhelming that the availability of alternative storylines in one's social network, including conspiracy theories, are welcomed for their ability to re-establish a sense of agency and control. A small proportion of publics actively denies the reality of anthropogenic global warming; nonetheless, a much larger proportion of publics accept the science of climate change but engage in *implicatory denial,* metaphorically putting our heads in the sand. For a growing number of us, however, neither denial nor apathy is available to us; the growing availability of knowledge about global warming and its implications is simply too difficult to ignore. Growing alarm about climate change can lead to anxiety, and for some, this may result in withdrawal. Individuals who have difficulty regulating emotions, are enduring other forms of distress, or simply lack a supportive social network, are more likely to withdraw. The coincidence of multiple crises—such as the pandemic, unemployment, racially-motivated violence—also have their toll on emotion coping capacity. Finally, many of us are very concerned about climate change, but are just *stuck*: at a loss as to how, or why, one should become involved in personal or collective efforts to address the climate emergency. For some this may be due to the lack of access to information, resources, or lifestyle change options. For others, inaction is supported by the lack of efficacy—a lack of confidence that personal and collective efforts to support change will make a difference.

Chapter 7 then turns to pathways supporting pro-climate action. While pro-climate action includes a wide array of personal and collective projects, any form of action will be associated with emotion-cognition pathways that depart significantly from inaction pathways. In this chapter, we explore the unique attributes associated with pro-climate action, by integrating the extensive but disparate empirical literature on conditions supporting agency in response to the climate emergency, including norms of responsibility; efficacy; and social belonging, which entails a return to important emotions like shame, pride and empathy. I will also discuss what I consider to be the importance of complexity thinking and future-gazing, two attributes highly associated with emotionality that come into play in our approaches to complex problems like global warming. Lastly, I discuss hope in detail, as perhaps the strongest motivator for pro-climate action.

Chapter 8, finally, discusses how can we apply this knowledge to facilitate the transformational changes required to support social and ecological wellbeing in our new climate reality. It's not over till it's over, as they say, and this is a race for our lives. This final chapter will begin with the premise that, based on what we know about human behaviour, we have the capacity to change, personally and collectively, and those actions can minimize the human and ecological toll of global warming, and maximize prospects for social wellbeing. The challenge: we need to strengthen mobilizing structures—those organizational elements that facilitate collective problem-solving—at a time when our emotional capacity to invest in such structures is weak. This chapter will be devoted to applying an emotionality lens to create strategies that have the potential to disrupt inaction pathways and sustain and upscale action pathways. As I argue, we can do so by attending to three basic human needs in our action projects: reflexivity, efficacy and belonging. While there are numerous possible ways of doing so, I discuss four key priorities to consider in our efforts to confront the climate emergency, each of which creates conditions for reflexivity, efficacy, and belonging. These priorities include cultivating empathy, envisioning futures worth fighting for, engaging in bottom-up politics, and most importantly, attending to emotions. In other words, feeling our way to change.

2 What Lies Ahead

It started with a map, and a warning. This animated map, showcased in many media sources towards the end of June 2021, showed a large, shape-shifting, amoeba-like dark red blob, shadowed in various red and orange hues, on a slow but determined course that would come to rest centred on top of Vancouver and Seattle, taking over 600 lives before it moved on. It ended with the complete obliteration over the course of mere hours of the small Village of Lytton, part of the traditional territory of the Nlaka'pamux Indigenous Nation, and home to 250 members of the Kanaka Bar Band. Lytton became world famous for about a minute for breaking a temperature record during that heat dome, right before the town was turned to ash. That record-breaking moment was followed so closely by the outbreak of wildfires on June 30 that residents had little warning. That tragedy should have ended there, more than enough for one region to bear, but it didn't, for the climate

DOI: 10.4324/9781003380900-2

emergency has no end in sight, no return to normal. In this very same region of southern British Columbia, while displaced survivors were still trying to figure out what's next, with grossly insufficient support from provincial and federal governments to support recovery, the 'Pineapple Express' arrived, a sequence of atmospheric rivers, one right after the other, dumping over 250 millimetres of water over a handful of days, water that landed on the same parched land that had been baked and then burnt just months before. Since the land no longer had any water absorption capacity, the water did as water does, it flowed, with such force that it ripped out highways and railways as if they were made of cardboard, and turned cities like Abbott and Merritt into twin Atlantises. With all transportation corridors to many small communities cut off, grocery store shelves quickly emptied. Electricity and clean drinking water were similarly in short supply. The TransCanada rail line, upon which Canada's natural resource-based economy is intimately dependent, stopped running for the first time since its completion in 1885. Proprietors lost money with every passing day that stores of grain and other exported goods sat in warehouses, while the shipping docks at the Port of Vancouver remained quiet. Dozens of families remain indefinitely displaced, as nearly every home in the Village of Lytton was completely destroyed by fire, and many more homes were destroyed by the flooding and fires throughout British Columbia.

These extreme events had the audacity to occur in Canada, one of the wealthiest nations in the world, which is why journalists were paying attention, but such disasters have befallen many other lands and communities across the globe, such as Mozambique and Pakistan, with much less fanfare. Events such as these, which are now occurring with mind-numbing regularity, are revealing—not in their occurrences but in their impacts—of the stark inequities that have been created by our neoliberal global economy.

I wish this were a movie, and soon the lights will come up, and we can shuffle out of the darkened theatre and back into the bright and comforting sunlight of normality. Alas, it is not, and we must begin this conversation about social responses to the climate emergency with an acknowledgement that the path forward will be fraught with economic disruptions, political upheaval, and trauma, along trajectories that, in the words of David Wallace-Wells,[1] have been deemed 'unseemly to consider.' Engaging collectively at the scale required to confront the climate emergency, while simultaneously coming to terms with substantial losses of lives, livelihoods, and things of value, will be formidable. What are the implications for each of us as individuals, and for civil society at large, of enduring anxiety-, loss- and trauma-filled days, months, years, at the same time at which the actions needed to meet this moment require personal and collective efforts akin to running an ultramarathon? With so many opportunities for fatalism and escape available to us, where will we find the motivation and capacity to simultaneously grieve, change our ways of living, and also confront a fossil-fuelled neoliberal capitalist economic system and its wealthy beneficiaries? The answer is to be

14   *What Lies Ahead*

found in the cultivation of the same human predispositions that have gotten us this far: our emotion-cognition predispositions that favour cooperation.

How Are We Doing?

The answers to this question run the gamut, with different data telling us different stories, but the only data that really matter to the planet are concentrations of greenhouse gas emissions and global average temperature trends. On both scores, the answer is: not good. In fact, really, terribly, frighteningly bad. Emissions levels to date have followed a path considered to be the 'worst-case scenario' in earlier climate models, depicted as RCP (Representative Concentration Pathway) 8.5. As of November 2023, the monthly average concentration of CO_2 in the atmosphere, as measured at the Maunakea Observatories in Hawaii (they used to be taken on Mauna Loa, until Mauna Loa volcano blew in 2022), was over 420 parts per million, a point along what appears to be a continuous climb since 1960. The trend line for methane, an even more potent greenhouse gas than CO_2, is even more alarming given its rapid uptick since 2010. Mirroring those greenhouse gas trend lines is the change in global average surface temperature.

These changes in atmospheric composition and temperature may not appear alarming—we are talking parts per million after all, and a degree or two! However, placed in historical context, they are quite alarming indeed. According to Luke Kemp and colleagues,[2] 'the current carbon pulse is occurring at an unprecedented geological speed and, by the end of the century, may surpass thresholds that triggered previous mass extinctions.' The last time

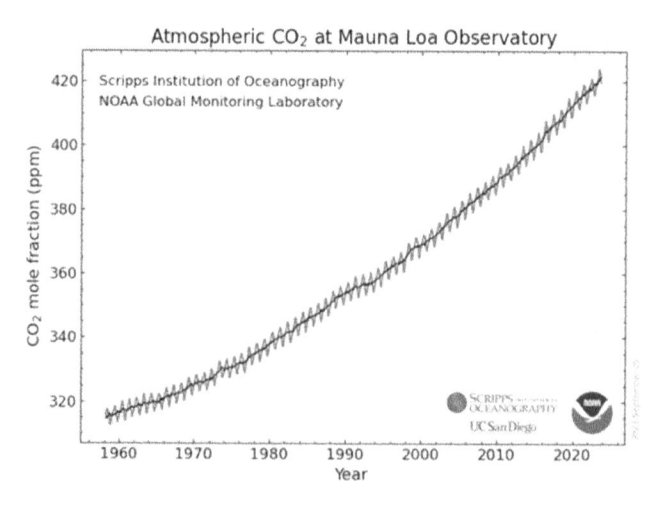

Figure 2.1a   Historical atmospheric CO_2 and global average temperature trend lines. Downloaded from https://gml.noaa.gov/ccgg/trends/, 5 September 2023; and https://www.climate.gov, 18 January 2023.

GLOBAL AVERAGE SURFACE TEMPERATURE

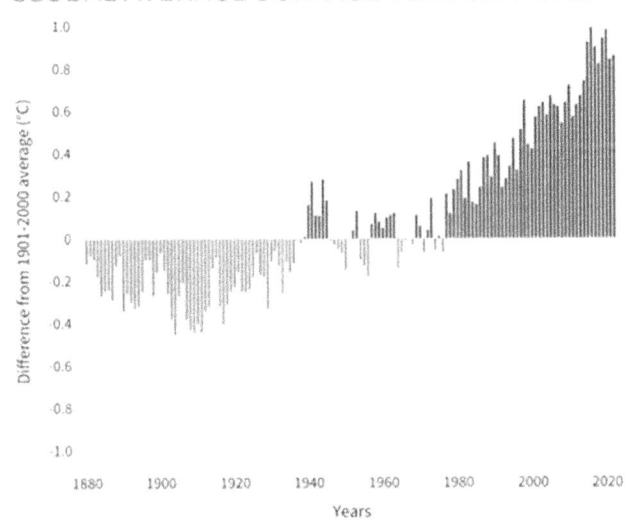

Figure 2.1b Continued

atmospheric CO_2 levels were this high was around three million years ago, a moment in earth's history when the global surface temperature was between 2.5 and 4 degrees Celsius above the average for the pre-industrial era, indicating that, by the time the planet adjusts to the current atmospheric composition, without any substantial interventions, temperatures will once again reach those levels. Of course, by then the concentration of greenhouse gases will be even higher, and the planet will be in a continuous catch-up mode, until those greenhouse gas concentrations are stabilized and diminished.

Global warming is now irrefutably on the radar of most citizens around the globe, and on the agenda of most nation-states, and most of these have made commitments to reduce their emission levels. There remains a rather large gap between what is needed and the ambition expressed by nation-states, but it should be acknowledged that the gap has been shrinking. On the other hand, most of those ambitions have not yet been *implemented*, and we cannot say for sure that all those commitments will ever be implemented fully—indeed, if recent history is any guide, the chances are quite slim. The authors of the IPCC Working Group I's Summary for Policymakers[3] captured these gaps in Figure 2.2.

Something else to keep in mind: efforts to reduce emissions will very likely be subject to declining returns on investment that characterize so many of our action plans. Trying to lose weight? Most likely, over the first few weeks you'll see a rapid drop on the scale, but each additional pound will seem more and more resistant to being disappeared. Initial efforts to reduce emissions can see quick, early rewards offered by low hanging fruit, like rapid adoption of increasingly affordable renewable energy technologies by

16 *What Lies Ahead*

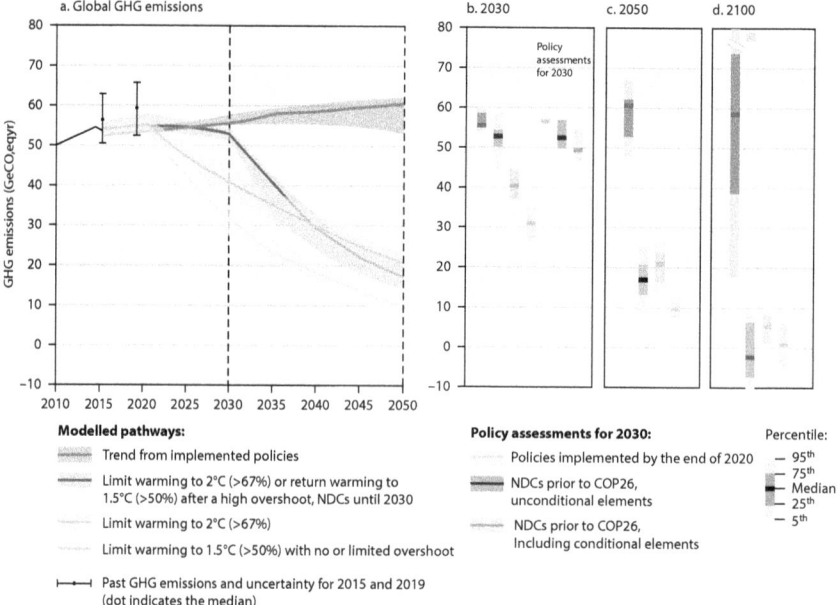

Projected global GHG emissions from NDCs announced prior to COP26 would make it likely that warming will exceed 1.5°C and also make it harder after 2030 to limit warming to below 2°C.

Figure 2.2 From IPCC Assessment Report 6, Working Group III, Summary for Policymakers (SPM.4), p. 19. This graphic is busy, but the key takeaways are offered, first, by the gap between the 'trend from implemented policies' line and the three lines below it indicating the emissions reductions needed to limit warming to 1.5 or 2 degrees Celsius, including a potential 'overshoot' trend line. The second key takeaway is gleaned by juxtaposing the left and right sides of the figure, namely the small gains that can be expected by 2030 if all current Nationally Determined Commitments are implemented. Baby steps to be sure, but at least they are in the right direction! Interested in digging deeper? Check out Fransen et al. 2023.[4]

those households and jurisdictions that have the infrastructural and financial capacity to do so. Chipping away at this gap further will get harder as we go, after that low hanging fruit has been picked.

One Very, Very Big Sticky Wicket: Fossil Fuels

Institutional change is hard. There are many, many features of modern institutions that must change in order to support low-carbon social systems, but the biggest sticky wicket is energy. Meeting the energy requirements of complex social systems is an enduring challenge to be sure, but that is not the main issue. The sticky wicket I want to highlight is the fossil fuel-dominated global energy sector, and the political and economic institutions that protect

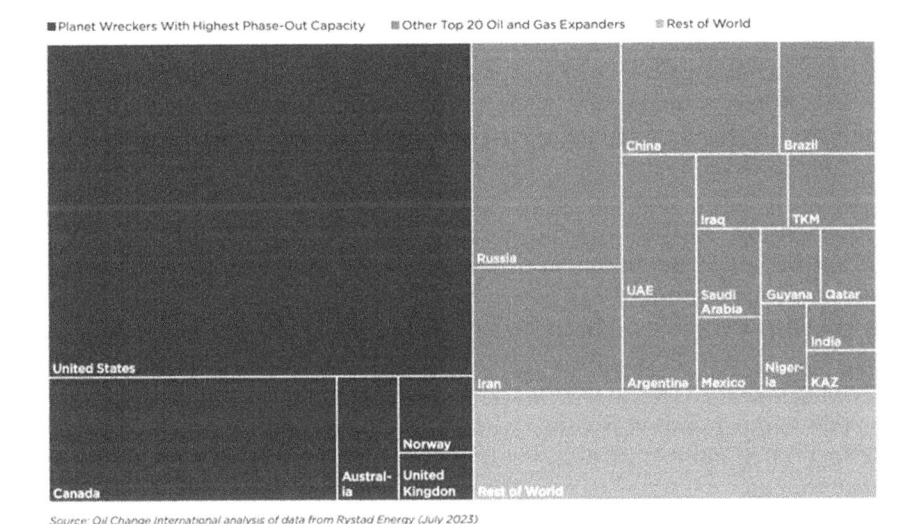

■ Planet Wreckers With Highest Phase-Out Capacity   ■ Other Top 20 Oil and Gas Expanders   ■ Rest of World

[Treemap showing countries: United States, Canada, Russia, Australia, Norway, United Kingdom, Iran, China, Iraq, UAE, Saudi Arabia, Brazil, TKM, Guyana, Qatar, India, Argentina, Mexico, Nigeria, KAZ, Rest of World]

Source: Oil Change International analysis of data from Rystad Energy (July 2023)

Figure 2.3 The elephants in the room. As featured in Planet Wreckers: How 20 countries' oil and gas extraction plans risk locking in climate chaos, by Oil Change International, September 2023, p. 5.

it. Emerging renewable energy technologies, and their growing appeal for policymakers, investors, and businesses, most certainly have begun to pose an existential threat to the fossil fuels industries, but those industries are not going to go quietly, and they may well do enough damage on their way out to render our transition ambitions moot. One recent analysis estimates that the current business plans of the world's 15 largest oil and gas companies will exceed their share of emissions that would be compatible with 1.5 degrees Celsius of global warming by over 100% by 2040.[5] These plans are being pursued with the full cooperation of many nation-states with fossil fuel reserves, most notably, you guessed it, the United States, followed by Canada. Oil Change International[6] recently conducted a tally of the fossil fuel production expansion plans among nation-states, painting a stark contrast between stated national climate commitments, and the relatively less advertised stated national fossil fuel production commitments, captured Figure 2.3.

Nation-state support also comes in the form of subsidies to the industry. The International Monetary Fund[7] estimated that as of 2022, subsidies to fossil fuel industries globally crossed the US$ 7 trillion mark, representing 7.1% of GDP that year, despite expressed commitments to eliminate 'inefficient' subsidies made by nation-states at COP26 in Glasgow. (Side note: don't ever let a politician, or your grandfather, convince you that we can't afford to address climate change! That is just bull pucky.)

A Snapshot of the Latest Climate Science

It's true, the future doesn't look *quite* as bad as it did even a few years ago—at least in terms of degrees of warming we can expect—given that many countries are finally directing their policy gaze towards mitigation, and the spectacular drop in the costs of renewables. This is good news, but needs to be eaten with a grain of salt, due to the number of caveats that must be accounted for, starting with the entrenchment of our fossil fuels sector and the low hanging fruit argument mentioned above. Another caveat is the fact that the accumulation of new research has led to a reduction in enthusiasm for the potential positive effects, or silver linings, of global warming, towards a growing agreement that the impacts to ecosystems and societies of a rapidly warming planet all sit squarely in the minus column, at least in the aggregate: the odd bumper crop in Russia's wheatbelt will not only be drowned out by the increasing number of bad years, but also by declines in productivity in other regions.

Among the most worrying of caveats, as remarked upon in several places throughout the Sixth Assessment Report from the IPCC[8] (which I rely on throughout this section unless otherwise noted), we are also learning that our planetary systems are far more sensitive to warming, with effects emerging sooner, and with more consequences, than once thought. In other words, even as gains are made in the form of mitigation commitments, we are discovering that the marginal benefits of those baby steps are lower than hoped. The following image, from Working Group II of that report, summarizes current assessment levels for five major Reasons for Concern (RFCs), showing a notable increase in comparison to earlier IPCC reports in both the levels of risk and confidence levels applied to these risk assessments. Say the authors of these findings:

> Compared to AR5 and SR15 [two previous major IPCC reports published within the last ten years], risks increase to high and very high levels at lower global warming levels for all five RFCs (high confidence), and transition ranges are assigned with greater confidence.
> (TS.C.12.1)

To make matters worse, researchers are learning that the warming potency of CO_2 actually *increases* with increasing temperatures[9]. In other words, one ton of greenhouse gases emitted today causes more warming than a ton emitted 20 years ago, and a ton emitted 20 years from now will cause more warming than a ton emitted today. I can relate! As my metabolism declines with age, every new calorie consumed becomes that much more effective at expanding my waistline. Of course my appetite thankfully has declined as well; not so our appetite for fossil fuels.

There is one more very big caveat we need to put on the table. Even realization of the most optimistic of mitigation scenarios—rapid reduction of

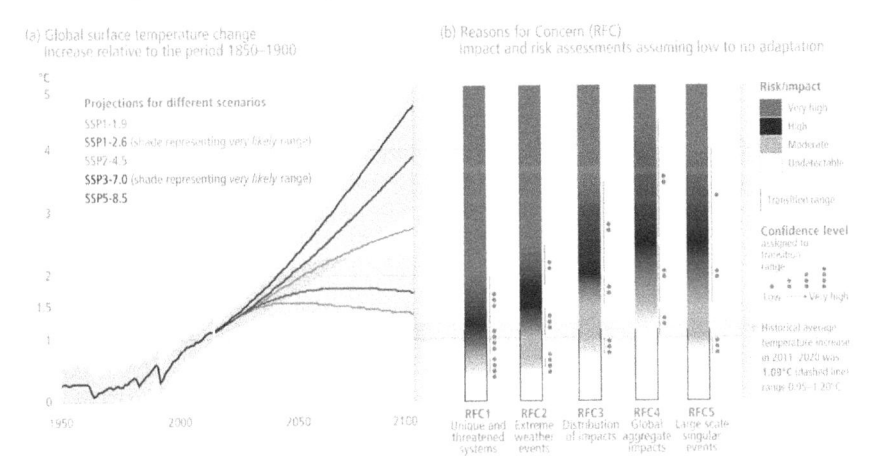

Figure 2.4 Summary assessment of risk levels in five Reasons for Concern. Figure SPM.3 in the latest IPCC Assessment Report, Working Group II Summary for Policymakers, p. 16. The figure indicates high to very high risk levels emerging well before two degrees of warming. In fact, we have reached the high-risk zone for two of them already.

emissions to limit warming to 1.5 degrees—will not erase the glaring inequities in the distribution of those impacts already baked in. The IPCC authors estimate that about 3.3 billion people are *currently* highly vulnerable to the impacts of climate change (TS.B.7.1); well over a third of the global population. But those highly vulnerable are not by any means randomly distributed across the planet. To be sure, direct exposure to global warming's impacts, like drought and sea level rise, varies by geography, introducing one important dimension to vulnerability's inequitable distribution. But to a far greater extent, the factors dividing the vulnerable from the resilient have more to do with what climate social scientists refer to as sensitivity and adaptive capacity. Outdoor trades workers are more sensitive to extreme heat, for example, and small, isolated rural communities have fewer resources to invest in adaptation. The biggest drivers of inequities in climate vulnerability, by far, are the same drivers of social inequity that have plagued us for 500 years: racism, sexism, colonialism, and an economic system designed to make the rich richer and the poor desperate. Many mitigation and adaptation efforts, in fact, run the risk of exacerbating inequity, when decisionmakers, operating from within these structures of inequity, do not make space for considerations of justice by, for example, choosing to invest in electric car infrastructure to the detriment of public transit. Or simply by reserving space at those decision-making tables only for the policy- and technological 'elite,' who also happen to be disproportionately anglophone, white, and male.

20   *What Lies Ahead*

Recent advances across the climate sciences are extensive—the escalation of investments in climate-related research is itself an important social response to the climate emergency—and I cannot by any means do justice to this accumulation of new knowledge here. Below I highlight just a few slices of this knowledge base, admittedly a slice that drew my personal, biased attention, and just a snapshot of those slices at best. There are now some very good, accessible sources of climate science for non-scientists available for those interested in digging deeper; I find websites like climate.nasa.gov and realclimate.org to be good places to start.

Extreme Events

This won't be news to anyone alive in 2023: the frequency and intensity of extreme events, those most vivid and visceral markers of global warming, are increasing with breathtaking speed, and have spawned a completely new set of terms to describe them—new to non-scientists at least—like bomb cyclones, heat domes, and atmospheric rivers. The Technical Summary for Working Group I, released before the wake-up call that was 2023, states that:

> Some extreme events have already emerged which exceeded projected global mean warming conditions for 2100, leading to abrupt changes in marine and terrestrial ecosystems (high confidence).
>
> (TS.B.2.2)

Escalations in three particular forms of extreme events have made themselves felt in recent years: fires, floods, and heat. Fires have ripped through North America, Southern Europe, Australia, and Russia with devastating speed and consequence. This chapter began with the story of just one of the recent fire events among many thousands experienced across the globe in the past five years, but a focus on Canada is apt. That fire occurred in 2021, a heavy Canadian fire season, but one that paled in comparison to the fire season that was 2023, turning Canada's vast forests from one of the globe's most reliable carbon sinks into a massive carbon source.

As with many climate-attributed extreme events, over the coming decades Canada's 2023 wildfire season may turn out to be less of an outlier than it appears today. By the end of the 21st century, analysts project that the area in the Boreal categorized as 'frequent fire-prone' will increase by 111%.[10]

The recent record of flood incidents is even more alarming than that for wildfires, if that is possible. Global warming is linked to increases in flooding due to overall increases in precipitation, the increase in frequency and intensity of extreme storms, and sea level rise, which produces king tides and storm surges with a much larger footprint than the actual rise in sea level. Over just the past few years, images and video footage of the sudden submersion of cities, communities, highways, and farms seem to drop faster than they can be absorbed, from California, where Central Valley residents

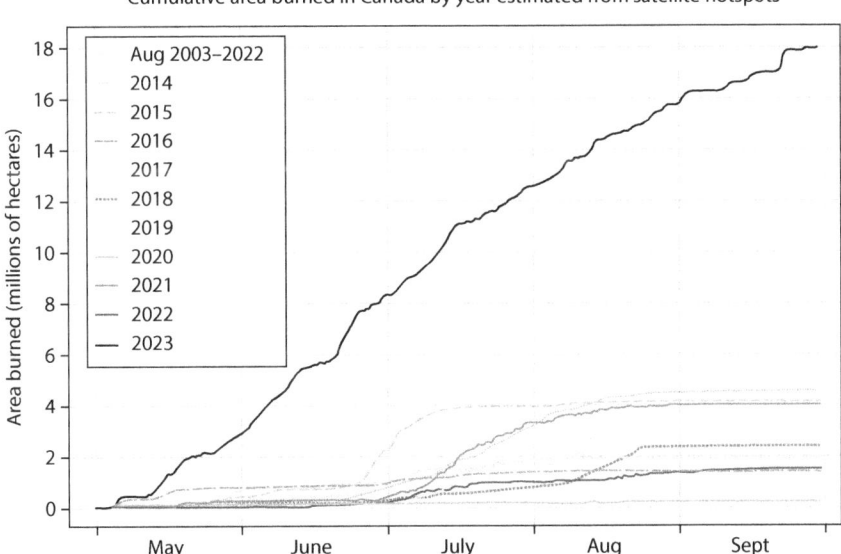

Cumulative area burned in Canada by year estimated from satellite hotspots

[Graph showing area burned (millions of hectares) from May through September, with lines for Aug 2003–2022, 2014, 2015, 2016, 2017, 2018, 2019, 2020, 2021, 2022, and 2023. The 2023 line rises dramatically to about 18 million hectares, far exceeding all other years.]

Source: Canadian Wildland Fire Information System

Figure 2.5 This graph by the paints a stark picture of just what a departure from 'normal' was the 2023 Canadian wildfire season, compiled months before that season came to an end. Source: Canadian Forest Service, Natural Resources Canada, reproduced with permission from the Department of Natural Resources, 2024. Available here: https://cwfis.cfs.nrcan.gc.ca/maps/fm3?type=arpt&year=2023

now have a lake where farmland used to be, to China, where cars are shown washing away like paper boats in the gutter. None of these quite reach the scale of the catastrophe that inundated a third (one third!!!) of Pakistan in 2022, killing an estimated 1,739 people, and causing nearly US$15 billion in direct damages, and another $15 billion (and counting) in lost economic productivity. Cities are particularly at risk given their cement-covered landscapes, and the location of many of our largest cities along coastlines and riverbanks.

As dramatic as floods and fires are, extreme heat affects a far higher number, and that number is projected to continue to grow. One recent study estimated that in 2021, the global population was exposed to a total of 3.7 billion more per capita heatwave days than the average for 1986–2005.[11] And while extreme heat events make a quieter entrance than bomb cyclones and fire tornadoes, they lead to the hospitalization and death of a greater number of people; people who are unable to escape its effects, like health-sensitive groups (children and the elderly in particular), outdoor workers, the houseless, and those who cannot afford air conditioning. In the U.S. alone, heat killed eight times more people than hurricanes, according to one recent news report.[12]

Water and Food Security

Threats to both water security and food security represent another silent killer that has been fuelled by global warming. According to the IPCC, half the world's population already experiences water scarcity for at least a month out of every year (AR6 WGII TS.B.4). There are many areas that are expected to see an increase in precipitation due to the warming climate, but that precipitation will not necessarily fall in places where it can be readily captured and delivered to thirsty communities, and can be cancelled out by warming surface temperatures, which enhance evapotranspiration. In my own province of Alberta in western Canada, most annual precipitation has historically fallen as snow, making ever so convenient natural water reservoirs in our mountains that slowly release their payload throughout the year. Even given an expected modest increase in precipitation here, however, with warming temperatures more of that precipitation falls as rain, not snow. Rather than being captured in our snowpack, it quickly enters our riverways and is carried out to sea before it can be captured. Albertans also join many other regions of the world in their reliance on glacial runoff, which will offer short-term bonuses with increased runoff under warming temperatures, until they are gone.

Shrinking water supplies have a direct effect on food security, a positive interactive effect that is anything but positive. When it comes to food production, however, water availability is just the tip of the iceberg, as the effects of global warming on agricultural productivity are myriad. Those extreme events wreak havoc, particularly drought, but also too much water-logging precipitation, storms that produce wind and ice damage, and outlier frost events all cause many a headache at farmer's tables. While weather has always been a dynamic variable that farmers are accustomed to, the increasing and broadening of types of variability are outside the bounds of conventional operational management. Farmers use historical records to determine the ideal time for seeding, for example, a date after which seedling-killing frosts are no longer expected, and they select crops on the basis of historical temperature records. Those historical records, however, are no longer a reliable planning tool.

Industrial farming efficiency is also becoming a source of increased vulnerability, and not solely due to large-scale monocropping. All plants have a temperature-to-productivity ratio, such that productivity is maximized at a certain temperature threshold, then rapidly declines when temperatures climb above that threshold. Those monocrop fields are planted in regions that are already close to the maximum threshold. In order to maximize productivity, in other words, a farmer will choose those crops and varietals that do particularly well in that region—wheat in Canada, olives in the Mediterranean, bananas in Panama, and corn in Iowa. But this also means those crops have no cushion; productivity begins to drop precipitously as temperatures increase. Heat stresses livestock as well, lowering milk and meat production, and raising fatality risks. Global warming plays havoc with the reproductive cycles of food crops as well. Changes in the timing of monsoons, for example, could mean that precipitation is no longer in sync with flowering for some perennials like nut trees; and pollinators may be scarce as well, with changes in the timing of insect emergence. Even a successful harvest, furthermore, cannot offer assurances of direct contributions to food security, due to the challenges that increased heat poses to food storage, enhanced conditions for food-borne disease pathogens like salmonella, and increased extreme weather compromising transportation routes.

All told, global agricultural productivity is set to decline precipitously even with adaptation (although adaptation can dampen the decline). One recent study,[13] focused just on South Asia, the most densely populated region in the world, estimates that every 1% increase in temperature produces a 1.93% decline in crop production.

Effects on Ecosystems

The many ways in which warming temperatures affect our ecosystems and their nonhuman inhabitants are alarming in and of itself, but global warming's effects on ecosystems also compromise human wellbeing, even if the pathways of causality are more indirect, and thus less noticeable to many. Shifting biomes means habitats are shifting faster than many species can migrate, particularly if development infrastructure stands in their way. This includes important food stocks like snow crabs off the coast of Alaska, which experienced what appears to be a sudden population collapse in response to warming water temperatures in recent years. Pathogens seem to be the exception to this rule, however, with disease vectors like zika virus and dengue seemingly capable of rapid migration with little standing in their way, cropping up in places with inhabitants—and health care systems—not previously exposed to these diseases.[14]

Disease vectors do not only affect people either: other species, including our food crops, are also increasingly exposed to pathogens migrating into new regions. While these bacterial, viral, and fungal species may enjoy expansion under warming conditions, prospects for many other species, and by extension biodiversity, are grim. Many human activities, particularly agriculture, have been hammering the habitats of other species for generations now, but climate change has exacerbated those processes to such an extent that researchers have come to view our current moment as the Sixth Mass Extinction, with extinction rates from 1,000 to 10,000 times higher than what would be expected without anthropogenic influence (If you haven't read it yet, Elizabeth Kolbert's Pulitzer-winning *The Sixth Extinction: An Unnatural History* is a must read!). Such losses to the earth's biodiversity have multiple implications for the planet, and for us, in ways we are just beginning to understand, but among the most obvious and concerning of implications is for our food supplies.

Emerging Signals of Climate Weirdness

The early 2020s has also been a time of alarming new scientific findings regarding the likelihood for abrupt shifts, or tipping points. Tipping points are difficult to capture in climate models, and hence have until recently largely been dismissed from climate science undertakings. However, many processes recently considered to be theoretical, low-probability, far-off-in-the-future possibilities have become empirical realities. In fact, a recent report states that at current levels of warming, we are at risk of breaching five global tipping points, and once we cross the 1.5°C barrier, three more will be added to the list[15].

A case in point is the effect of global warming on the Atlantic Meridional Overturning Circulation (AMOC), the giant conveyor belt that redistributes cold water from the north, and warm water from the south, in the Atlantic Ocean. Long understood to play a very large role in maintaining the temperate climate of Europe, among other things, the theory that global warming might slow and perhaps even shut down AMOC has been noted, but largely dismissed as a low-probability shift not likely to be observed over the course of this century. Then it happened—signs of slowing can now be detected, forcing researchers to revise those projections sharply upward, with one recent analysis projecting a collapse by 2050 under current emission scenarios.[16]

The planet's ice cover has also offered up some surprises for climate scientists in recent years, including previously unimaginable heat waves occurring near the North and South Poles. Researchers are also learning that ice sheets can collapse much more rapidly than previously thought. Once considered to be less sensitive to climate change than other regions, at least in the near term, Antarctica in 2023 experienced the lowest sea ice extent on record, and the rate of observed warming in the Antarctic in recent years is well above the warming projected by climate models—an observation generating growing concern that our models may be underestimating other impacts of global warming as well. The focus of much of Antarctic research is centred in particular on the Thwaites Glacier, also called the Doomsday Glacier, since, at 192,000 square kilometres across and around a kilometre deep, it holds enough water to produce over two feet of sea level rise all on its own. The glacier itself, and the ice sheet holding this glacier in place, however, are beginning to show signs of weakening in the form of cracks and fissures, suggesting the potential of an abrupt collapse.

The Spill-over Effects: Social Consequences Beginning to Unfold

The direct impacts of global warming—to water levels, hurricane occurrence, crop production—are not exactly *easy* to detect with our current toolbox of scientific methods; they involve high levels of complexity and require sophisticated measurement techniques and no small amount of research effort. However, once we begin to turn our sight to the spill-over effects as these impacts infiltrate our political, cultural, and economic systems, that level of complexity gets ramped up several notches. (And yet, the level of investments in research in the climate *social* sciences, unfortunately, is paltry by comparison to our natural science counterparts. Don't get me started on that tangent.) This is nonetheless the domain I wish to focus on here: the social effects of unfolding climate impacts, beginning with those most acutely felt, the social consequences of disasters.

The Social Consequences of Disasters

The costs of disaster have taken a turn towards the eye popping since 2020. The numbers of deaths are orders of magnitude higher in low-income countries due to more limited emergency response capacities, while the economic costs show the reverse trend, in the form of property damage, lost income, and lost business much higher in wealthier countries. Tallies of economic costs counted in dollars do not capture the relative economic consequences of those material impacts, however, to families, communities, and governments. The consequences of such losses are much more acute in low-income countries, and for poor and under-served communities and families everywhere. The loss of an insured, million-dollar vacation home—which registers as a substantially larger impact than the loss of a shantytown—may ultimately have little to no effect on the material wellbeing of a wealthy family, but the loss of those shantytown homes, or even an uninsured home in a middle-class neighbourhood in Colorado, is devastating, even if the value of those homes is miniscule by comparison. Insurance coverage is higher among wealthier families; disaster relief is greater in wealthier nations; the social support networks of those in middle- and upper-income strata have much deeper pockets. On the other hand, even though the costs of disaster are most often counted in dollars and fatalities, the repercussions for mental health and human wellbeing, for community, and for society are far more extensive, and do not necessarily correlate with the level of material consequences. There is no question that direct experience with disaster compromises mental health, with some survivors experiencing depression, and post-traumatic stress disorder long after the event, across the economic spectrum.

The potential for conflicts and violence to erupt in the aftermath of disaster can exacerbate these mental health effects, and compromise recovery efforts. Studies of social responses to disaster often celebrate the degree of care and cooperation that many community members express in the face of disaster, a sort of silver lining to the enormous costs incurred. Rebecca Solnit[17] offers several accounts of altruism and cooperation in the aftermath of disaster that survivors later recall with nostalgia. This outcome is far from universal, however. Even in the absence of extreme events, the gradual shifts in temperature and precipitation caused by global warming have been found to increase the risk of human conflict.[18]

Additional not so positive features of disaster response need to be noted. The first is the underlying inequity in recovery capacity, and the second is the decline in resilience that can be expected with increased disaster regularity. One layer of inequity has already been alluded to: Some survivors have insurance coverage; others have none. Some have cars to evacuate in and cash reserves for temporary housing; others do not. Some get paid leave from their employers; others lose their jobs. Other dimensions of inequity are less obvious but no less prominent. The most important source of support for survivors, more than anything else, are informal networks—family and friends. The coping capacities of newcomers, and individuals who for whatever reason are more socially isolated than others, are thus much lower. Then there are the differences imposed by intersectional positionality. Marginalized groups not only pay relatively steeper economic costs, they are also more susceptible to negative physical and mental health outcomes.[19] Women and girls, non-white residents, and those with lower education and more precarious employment have been shown in numerous studies to have lower overall resilience in response to disasters.[20] One recent analysis of U.S. residents found that Black and Hispanic residents were more than 2.5 times more likely to fall into the highest impact group in response to extreme events, with women and low-income households also significantly more likely to be highly impacted.[21] Marginalization features in numerous forms of second-order effects as well. To take one heart-wrenching example, increases in child, early, and forced marriage post-disaster have been observed.[22]

In fact, disasters can exacerbate inequalities, as the more privileged experience full recovery, while the less privileged find themselves worse off, resulting in an even larger gap between the haves and have-nots.[23] Not surprisingly, studies show that households that rely upon natural resources for their livelihood are especially vulnerable to financial losses.[24] But the inequities do not stop there. Those haves simply have the political clout to capture far more than their share of relief funds, despite the fact that they need it the least. Several analyses of relief fund distribution after Hurricane Katrina are telling, with funds being funnelled disproportionately to large corporations and more lightly damaged but wealthier neighbourhoods.[25] Redevelopment decision-making all too often favours business interests as well.[26] Naomi Klein calls this trend *disaster capitalism*,[27] in which such tragedies become opportunities for profit-making, which certain companies are well-situated to capitalize upon. These processes of inequality are not just about the capabilities of the privileged, however. They also reflect a systemic regard for some survivors, and disregard for others, like Indigenous communities, which seem to always be on the front lines of extreme events, and the last in line for recovery assistance, in my own country.

Least studied but very worrisome among social consequences in the face of the climate emergency is the prospect for resilience to decline over time with repeated disaster exposure. Disaster experience, as with all other forms of experience, offers learning moments. These lessons can build adaptive capacity, or resilience, that can enhance preparedness for future disasters.

But all disasters, whether or not lessons are learnt, require a certain amount of recovery time—time to rebuild, time to replenish, time to heal. That time frame can be quite long indeed, raising the possibility that, in some places at least, the very notion of full recovery is a misnomer. The physical scars of Hurricane Katrina can still be seen in some neighbourhoods nearly 20 years later. Emotional scars may never fully heal. That is precisely what Kai Erikson[28] found, working with the survivors of the Buffalo Creek flood.

But what happens when the once in a lifetime event becomes the once in a decade event? Or once a year? Once considered Events of the Century, the fuel that global warming has added to extreme events like heat domes, ice storms, typhoons, and wildfires means that they occur with increasing regularity, and affect more of us, sometimes with multiple disasters erupting simultaneously, as with the heat and fires, followed by floods that hit southern British Columbia in 2021. A whopping 86% of U.S. respondents to a recent survey[29] reported being affected by extreme weather in one way or another, and 34% indicated that they were very or extremely impacted. Research on the cumulative effects of repeated traumatization for individuals, families, and communities is miniscule, so we just don't know what those effects look like. Even studies that have been done to understand the personal impacts of exposure to singular traumatic events provide mixed signals, with some indicating prior disaster experience can serve as a buffer against potential future events, while others suggest prior experience instead renders survivors more sensitive to future events.[30]

Declines in resilience are not solely an individual-level phenomenon either. Given the compounding effects of collective trauma, I am left wondering about thresholds—social tipping points—below which communities simply run out of the steam required to pick up the pieces. Many studies of trauma focus on the individual experience, say the soldier returning home. These studies consistently show that the most important ingredient for recovery is the support offered by one's family, community, and social networks. In other words, according to this scenario, the trauma survivor recovers by virtue of their embeddedness in a community of individuals who are capable of offering emotional support. Collective trauma is something different entirely, in which there are far more in need of receiving such support than there are individuals with the capacity to give it. Communities with strong ties do tend to experience improved recovery in comparison to those that do not.[31] However, those community social support networks can themselves be destroyed in the aftermath of disaster.[32]

Migration

One response to disaster, and also the slow-moving shifts in temperature and precipitation regimes that render certain places unlivable, is migration. Massive movements of people seeking better living conditions have already become a hot-button political issue, particularly for Europe, the destination for many migrants from African nations, and for the United States, the destination for migrants from Central and South America. The goodwill of many of these receiving nations has seemingly already been breached, with growing anti-immigrant rhetoric, especially among extreme right-wing factions. As of Fall 2023, even U.S. President and migrant sympathizer Joe Biden has succumbed to calls to expand the border wall between the United States and Mexico, a policy once considered anathema to any but hard-core Trump supporters.

Most migration occurs *within* countries, however, and thus is less likely to make it to the front pages of mainstream media in the North, but nonetheless is of concern. These migrant flows almost always move from rural regions to rapidly growing cities across the Global South; cities that can hardly be considered climate safe havens. According to the IPCC AR6 authors' best estimates, the world's cities will be confronted with an additional 2.5 billion residents by 2050, with most of this increase in Asia and Africa (TS.C.9.1). Many of these migrants land in informal settlements, already home to around a billion inhabitants, located in precarious settings like landslide-prone hillsides, and lacking in basic services like electricity, water, and sewage services. Many of these cities, ironically, are located along climate-risky coastlines and thus are likely themselves places from which residents may inevitably be forced to flee in the future. Authors in the IPCC's Working Group II have projected over 100 million people in Africa alone will be exposed to sea level rise by 2030, and in another 20 years, over a billion people globally will be at risk (TS.C.5.2). Not all of those people are necessarily in situations that would prompt migration. But neither are coastal regions the only places where climate risks raise the spectre of displacement: those regions already facing the most severe levels of food and water scarcity are the original homes of many migrants on the move today. The Institute for Economics and Peace[33] projects the displacement of over one billion people by 2050, a number projected not on the basis of climate impacts alone, but on their disproportionate severity in precisely those countries that are already enduring high levels of economic precarity and political conflict.

Migration is extremely dangerous for the migrants themselves, as the many lives lost in the Mediterranean Sea and along the Darien Gap illustrate. But even success stories have social costs. Male income earners are most often the ones to leave (although many families, and unattended children, also migrate), and this splintering of family units takes an emotional toll on both the leavers, as migrants reside in places in which they have little sense of belonging, and on those left behind, in communities in which human capital is drained.[34] As dangerous as migration is, the migrants are the lucky ones, in comparison to those too poor to leave. This may be changing, however: One recent study even suggests that the negative economic impacts of global warming may inhibit mobility as much as it prompts it, as the limited economic resources that are pooled to support migrants shrink.[35] Combined with the removal of welcome mats in many receiving countries, a growing number of climate refugees will be left with no place to go; no choice but to attempt to survive in places no longer fit for survival.

Cumulative Economic Consequences

The economic consequences of global warming are not restricted to those already surviving at the economic margins. These consequences come in many forms, often dispersed and difficult to assess. Increases in property insurance and lowering of property values are new realities for many in disaster-prone regions. So are sporadic spikes in the prices of certain foodstuffs due to harvest failures, from rice to olive oil. Spikes in disaster relief payouts are threatening the viability of the rainy-day funds of even wealthy nations like the U.S., indicating the possibility that as the economic costs to households and communities increase over time, social welfare coffers will be running dry. And, we have begun to realize the enormous costs imposed by global warming to labour productivity, with its effects scaling up through entire economic sectors and regions. Heat led to an estimated loss of 470 billion labour hours globally in 2021 alone, culminating in a 5–6% hit to the GDP of those countries that can least afford it.[36] Such declines in labour productivity have a negative impact on business, and also for families: some of those labourers may have the benefit of paid leave, but most do not.

Health

The health consequences of living in the climate emergency are becoming more and more clear with each passing year of new research. A central takeaway from this accumulating research record is the sheer number of ways in which global warming implicates human health, including the acute impacts of extreme events, declining access to healthy food and clean drinking water, the growing prevalence of infectious disease vectors and their expansion into regions not previously exposed to those vectors, and their aggravation by climatic hazards such as floodwater.[37] As should come as no surprise by now, these health consequences emerge along the same inequitable pathways as economic consequences, due in part to the reduced access to health care, and in part to the higher prevalence of pre-existing poor health conditions in under-served communities. Nina Hall and Lucy Crosby[38], for example, found that Indigenous Australians are almost twice as likely than settlers to have a disability or chronic disease and have markedly reduced life expectancy, about 10 years less than settlers.

As already discussed, the health consequences of the climate emergency include mental health. Poor mental health emerges most acutely in response to extreme events,[39] but even the anticipation of such experiences, and contemplation of the multiple other impacts of global warming, induce anxiety.[40] This climate anxiety is reaching epidemic levels among youth in particular,[41] and also for those whose livelihoods are particularly sensitive to the impacts of global warming, like farmers. In India, for example, global warming played at least some role in the suicides of as many as 60,000 farmers over three decades.[42]

Declines in Quality of Life

The World Health Organization projects global warming will cause an additional 250,000 deaths per year between 2030 and 2050, while another study estimates the loss of one billion lives by 2100, if we reach two degrees of warming by then.[43] These numbers should draw our attention and concern. However, as should be clear by now, all such future projections should be treated with caution, and I think this is true for fatality estimates in particular—those numbers could be reduced considerably depending on the actions taken to prevent them; they could also be much higher.

We will be affected by warming in many ways that are not marked by death, however. Each of the social consequences described above—extreme events, economic decline, and compromises to mental health—will affect many but by no means all, given the enormous disparities in vulnerability. But, no amount of wealth, privilege, vitality, or luck can ensure protection from the compromises to quality of life already unfolding as a consequence of our rapidly warming home. Already a part of the lives of everyone in my social circle are smoke days leading to cancellations of school and outdoor activities; evacuations leading to displacement for extended periods of time; strict water rationing in some cities; relentless, cranky-making heat that contributes to increases in hostility and violence; ecological grief for the loss of loved naturescapes and more-than-human beings, leading to a feeling of homelessness at home[44]; and summer holidays—a treat that for many in the middle classes is one of the few bright spots in a year of overwork and multitasking—that go very, very wrong.

The space for water-based recreation is shrinking across the Western U.S., where Lake Oroville, along with several other water bodies in the region, has become a bath tub, possibly permanently, compromising the recreational opportunities of the wealthy and working classes alike. Photo taken on May 2021. Source: Wikipedia.

Collective Action!

On the upside, we have experienced an impressive rise in ecologically oriented social movements across the globe. While I have spoken to many an activist who feels disappointed at the seeming nothing burger-ness of a single protest action, the cumulative effect of those actions cannot be ignored. As someone who has been observing environmental movement activities for decades, I can safely say that we are experiencing an impressive, and very encouraging, upscaling of the climate movement, with new factions like Fridays for the Future, Extinction Rebellion, #landback, and the Sunrise Movement, making headway from the floor of the United Nations Assembly to the courtroom.

More difficult to detect but just as noteworthy are the shifts in discourse that parallel this movement activity, with previously dismissed paradigms—paradigms that can provide important foundations for constructive social change—growing in prominence and legitimacy. These include serious consideration for dematerialization, degrowth, and new respect for the agency of, and our relations with, nonhuman beings.

What Lies Ahead. Maybe

Conventional science, including climate science, typically involves dismissing system features that are either unquantifiable or associated with very low probabilities—the tails of our data distributions—even if those low probability outcomes may entail large consequences. This includes what risk managers refer to as worst-case scenarios, which, in the case of global warming, entail probabilities that are not all that small. Some, in fact, are worryingly large. Take the prospects for shooting past two degrees of warming. Even though the current estimated probability of doing so is well above zero—more likely than not by some estimates—most researcher attention on global warming's impacts is focused on scenarios between 1.5 and 2 degrees. Consideration of what might unfold if we shoot *well past* two degrees, bringing to mind images of full-scale collapse, is rarer still. The exceptions here prove the rule—the handful of scientific analyses, including in particular the work of James Hansen and colleagues, most recently represented in their paper 'Global warming in the pipeline,'[45] have been heavily criticized by peers. To find further serious contemplation of such low-but-not-zero-probability outcomes, readers need to move into writings outside the bounds of academia, including, for example, the work of journalists like David Wallace Wells, intellectuals who have left academia like Jem Bendell, and of course, an explosion of cli-fi, some of which feels a bit too on the mark (Robinson's *Ministry for the Future* is not to be missed).

Such scientific practices of conservatism are not reflective of negligence on the part of the scientist. To the contrary, conservatism is prescribed by the scientific method. But that method, set in stone long before scientists were grappling with anything quite as messy as anthropogenic global warming, is not designed for such complex systems, in which those dismissed outliers can be a very, very big problem. Worst-case scenarios do happen, all the time, and

32 *What Lies Ahead*

the less we conceive of their plausibility, the less we are prepared. Feedback effects in our planetary systems are confounding enough, but what gets lost in particular are the emergent outcomes that result from the myriad social phenomena that intersect with those planetary processes. This means that the chances are very good that many of these social consequences, even the purportedly quantifiable economic consequences, are grossly underestimated.[46] Below I peek at an admittedly selective case of possible social outcomes of concern, each representing potential social tipping points that can lead to the eventual collapse of modern civilizations, particularly if they surface in combination. Are these scenarios realistic? Maybe, maybe not. But they certainly cannot be ruled out. In the words of Isabelle Stengers,[47] having handed over the reins of society to capitalists, 'we are exceptionally ill-equipped to deal with what is in the process of happening.'

Economic Disintegration

The costs of mitigation and adaptation today are large, but they pale in comparison to the eventual costs of doing nothing. More money spent on basic needs at the household level, and on disaster response at the governmental level, translate into less investable or spendable income. This does not simply foretell declines in economic growth, they suggest a potentially drastic shrinking of the economic pie. Infusions of disaster relief and other forms of aid have been important to maintaining stability across the globe, and crucial to low-income countries suffering the scars of decades of colonialism,[48] but such forms of aid are unlikely to continue to flow in the face of repeated disasters and multiplying disaster zones. When those coffers run dry, many families, communities, and entire nations will be unable to rebuild. These losses, in the aggregate, will affect global economies. These disruptions—that just feels like an inadequate word—look like destitute families, the failure of businesses, banks, entire industry sectors, and currencies gone haywire.

Institutional Failure

We may already be here. Western social institutions emerged amidst sociomaterial configurations that simply no longer exist; we now face a jarring disjuncture between institution and reality, between endogenous structure and exogenous forces. Notably, those institutions we rely upon for accommodating, moving through, loss have come up woefully short. The insurance industry is already experiencing a reckoning, with the residents of many disaster-prone regions unable to secure insurance at all. The fragility of health care has also been brought to light in the face of the COVID-19 pandemic. Citizen confidence in the continued support of their governments is being put to the test in many places around the globe. Confidence in intergovernmental institutions like the United Nations has fared even worse, as the record of success at tackling global crises like global warming and the pandemic, and mediating against geopolitical conflagrations like Russia's invasion of Ukraine and Israel's pummeling of Palestine, is woefully disappointing. Threats to the functionality, and legitimacy, of institutions are further hampered by *polycrises*: the simultaneous occurrence of independent crises—such as pandemics, wars, and harvest failures occurring all at once, that can, as Kemp and colleagues[49] speculate, erupt into 'system-wide, synchronous failures.' As the institutions that offered a semblance of stability during modernity begin to lose their warm blanket of legitimacy, that's when disruption really begins to take hold.

Climate Barbarism

Tragedies can bring out the best in us; they can also bring out the worst. Experiencing the impacts of the climate emergency has already been linked to increases in domestic violence, including against intimate partners[50], and children.[51] Even exposure to information about global warming, according to one recent study,[52] has been found to generate higher levels of threat perception towards out-groups. Emerging signals of anti-egalitarianism, extreme right-wing populism, xenophobia, white supremacy, competitive hyper-localism, and democracies on the chopping blocks have given researchers such as Naomi Klein[53] and Jacob Blumenfeld[54] reason to contemplate the realistic probability of an emerging *climate barbarism*—or effectively, the elimination of the civil from civil society. According to Klein,[55]

> Climate barbarism is a form of climate adaptation. It is no longer denying that we have begun an age of massive disruption, that many hundreds of millions of people are going to be forced from their homelands, and that huge swathes of the planet are going to be uninhabitable. And then, in response to that, rather than doing all the things that are encoded in the UN Convention on Climate Change, which recognizes the historical responsibility of many of the countries that happen to have a little more time to deal with the impacts of climate change—are insulated both by geography and relative wealth—instead says, look, we simply believe we are better, because of our citizenship, because of our whiteness, and our Christian-ness, and we are locking down, protecting our own, pulling aid.

Depopulation

Many forms of blame for the climate emergency have surfaced, some of which are entirely warranted. Those countries that industrialized first, for example, have received a heavy dose of blame, given that they filled their coffers with the help of fossil fuels, and their historical share of emissions is substantial. Entirely warranted, but not always helpful when such blaming becomes justification: we deserve our slice of the atmosphere too, despite the fact that there are just a few thin slices and crumbs left.

34 *What Lies Ahead*

On the entirely unwarranted side, on the other hand, no culprit gets more attention than us. Specifically, our numbers. Overpopulation is the go-to punching bag for many a climate advocate. To be clear, 'overpopulation'—and it has yet to be scientifically determined what level of global human population constitutes the 'over' threshold—did not *cause* climate change, or any other environmental malady. This storyline is just pure racism cloaked in some bastardized version of 'ecology,' a deflection of accountability from among the most privileged, and most accountable.

On the other hand, an aversion to any discussion of population at all is, albeit guided by the best of moral intentions, equally folly. We do need to talk about population. I know, as a feminist and anti-colonialist, raising the subject of population is anathema. It is absolutely true that ecological footprints are vastly unequal, and curtailing the profligate consumption and waste of the few would have a substantial positive benefit, and targeting the wealthy and their disproportionate emissions ought to be a policy priority. But the fact remains that our climatological and ecological destruction has been shrinking the carrying capacity of the earth to support humans, and all other living beings, and this absolutely compromises our aspirational sustainable development goals of providing for all eight—soon to be nine, and then ten—billion of us. Basic needs like land, freshwater, and food are already limited in many regions as a result of their commodification, and becoming far more so due to the impact of climate change. So, the Great Acceleration in material consumption and waste is crashing into the Great Shrinking of planetary carrying capacity. Our current mismatch between population size and carrying capacity is NOT about women's reproduction. Ironically, with the notable exception of China's population control policy, which they are desperately attempting to reverse now, control over women's reproductivity has almost always translated into increases, not decreases, in fertility, as women's right to contraceptives, their rights to occupations outside the home, their rights period, have been shackled by their patriarchal keepers. It is also NOT about the supposed uncontrollable selfish greed of the human race, as some Anthropocene narratives would have it. It is about the massive-scale terraforming of the planet and its skies by a small minority of privileged, colonial interests over the course of the past 600 years.

The result? Many people will die. Just how many is not a prediction that is easy to assess, although some have tried. So, yes, population is something we need to discuss: namely the prospect of rapid *de*population in many regions of the globe, and its consequences for economies, communities, and our collective conscience.

Eyes Wide Open

Always a Better Approach than Heads in the Sand

Yikes, I'm depressed after writing this chapter. And I haven't yet mentioned what I find to be the worst possible outcome. For me, this looks like

Normalization. A collective forfeiture before the game clock expires. A form of playing dead in the face of threat, but a walking, living, breathing dead. The death of hope, the death of personal investments in collective survival. This normalization can already be observed, among those who can afford it. For the top 1%, it is observed in the construction of luxury bunkers in New Zealand, or white masculinist fantasies of colonizing Mars. For the top 20%, it looks like mindfulness retreats and virtual reality games. For many others, it looks like denial (*my* life won't be affected. It won't be that bad. Technology will save us!) and withdrawal (wake me up when it's over!), both to be taken up in a later chapter.

If I thought all bets were off, I would not be putting in the effort to write this book. None of the social outcomes laid out above is absolutely foretold. Complex systems are replete with uncertainties, and unexpected interactions. No one foresaw the rapid decline in the cost of renewables experienced in just the past ten years; no one foresaw the stimulus the Russian invasion of Ukraine provided to energy transition across Europe. And no one saw the Greta effect coming, from the modest beginnings of a one-person protest to a global youth-led movement. We do not need to place our faith solely upon such Black Swans either. We, as a species, do have inherent capacities that I believe provide us with the capability to confront the climate emergency, particularly our capacities for cooperation.

But those inherent capacities have been for decades sanctioned out of us by a Western culture that has become entirely beholden to capitalism. And so, it is Humanity V. Capitalism. The utopian claims that capitalism spurs progress, raises all boats, and serves the interests of everyone can at this point—actually decades ago and at least at the time of the 2008 financial crisis—be removed from our lexicon of legitimate claims to reality. Capitalism definitely spurs inventiveness, in the pursuit of greed, along a route entirely devoid of moral compass. As mentioned earlier, even catastrophe is, with enough creativity, an opportunity for profit. Capitalism has also normalized massive poverty and planetary destruction. The good news is, the turn of the 21st century has been a time of bursting bubbles, not just of real estate markets which is by no means good news for indebted homeowners, but of our collective euphoric belief in the myth of capitalism. But with no viable resistance, we have come to simply hold our noses, and proclaim that climate protestors are the real nuisance and not the kilometres of flooded and burnt landscapes.

A nice, happy-ending family movie sounds very attractive right now. But real, meaningful change begins with reality checks. It begins when we remove our blinders and face crises with our eyes wide open. You may think it's bad now; but it's going to get much, much worse. Bank accounts will evaporate. Careers will end. Species, and entire ecosystems, will be lost forever. Food supplies will shrink. People will die, far more than have already, and the vast majority will be among the poorest of the poor—in other words those least responsible for generating the crisis in the first place. Our capacity to care will be taxed to the limit.

We need to take a good hard look at that very real scenario, so we can build a new one to take its place.

Notes

1. Wallace-Wells, *The Uninhabitable Earth*. P. 9.
2. Kemp et al., "Climate Endgame."
3. Intergovernmental Panel On Climate Change, *Climate Change 2021—The Physical Science Basis*.
4. Fransen et al., "Taking Stock of the Implementation Gap in Climate Policy."
5. Romanello et al., "The 2022 Report of the Lancet Countdown on Health and Climate Change."
6. Oil Change International, "Planet Wreckers: How 20 Countries' Oil and Gas Extraction Plans Risk Locking in Climate Chaos."
7. Black, Parry, and Vernon, "Fossil Fuel Subsidies Surged to Record $7 Trillion."
8. Intergovernmental Panel On Climate Change, *Climate Change 2022—Impacts, Adaptation and Vulnerability*; Intergovernmental Panel On Climate Change, *Climate Change 2021—The Physical Science Basis*.
9. He et al., "State Dependence of CO_2 Forcing and Its Implications for Climate Sensitivity."
10. Senande-Rivera, Insua-Costa, and Miguez-Macho, "Spatial and Temporal Expansion of Global Wildland Fire Activity in Response to Climate Change."
11. Romanello et al., "The 2022 Report of the Lancet Countdown on Health and Climate Change."
12. Milman, "'Silent Killer': Experts Warn of Record US Deaths from Extreme Heat."
13. Chandio et al., "Climate Change and Food Security of South Asia."
14. Van Wyk, Eisenberg, and Brouwer, "Long-Term Projections of the Impacts of Warming Temperatures on Zika and Dengue Risk in Four Brazilian Cities Using a Temperature-Dependent Basic Reproduction Number."
15. Lenton et al., "The Global Tipping Points Report."
16. Ditlevsen and Ditlevsen, "Warning of a Forthcoming Collapse of the Atlantic Meridional Overturning Circulation."
17. Solnit, *A Paradise Built in Hell*.
18. Hsiang, Burke, and Miguel, "Quantifying the Influence of Climate on Human Conflict."
19. Arcaya, Raker, and Waters, "The Social Consequences of Disasters."
20. Chen et al., "Anxiety and Resilience in the Face of Natural Disasters Associated with Climate Change."
21. Zanocco, Flora, and Boudet, "Disparities in Self-Reported Extreme Weather Impacts by Race, Ethnicity, and Income in the United States."
22. Doherty, Rao, and Radney, "Association between Child, Early, and Forced Marriage and Extreme Weather Events."
23. Williamson, McCordic, and Doberstein, "The Compounding Impacts of Cyclone Idai and Their Implications for Urban Inequality."
24. De Silva and Kawasaki, "Socioeconomic Vulnerability to Disaster Risk."
25. Adams, "The Other Road to Serfdom"; Gotham, "Reinforcing Inequalities"; Gotham, "Re-Anchoring Capital in Disaster-Devastated Spaces"; Arcaya, Raker, and Waters, "The Social Consequences of Disasters."
26. Gotham and Greenberg, *Crisis Cities*.
27. Klein, *The Shock Doctrine: The Rise of Disaster Capitalism*.

28 Erikson, *Everything in Its Path*.
29 Zanocco, Flora, and Boudet, "Disparities in Self-Reported Extreme Weather Impacts by Race, Ethnicity, and Income in the United States."
30 Chen et al., "Anxiety and Resilience in the Face of Natural Disasters Associated with Climate Change."
31 Wooten, *We Shall Not Be Moved*; Aldrich and Meyer, "Social Capital and Community Resilience."
32 Arcaya, Raker, and Waters, "The Social Consequences of Disasters."
33 Institute for Economics and Peace, "Over One Billion People at Threat of Being Displaced by 2050 Due to Environmental Change, Conflict and Civil Unrest."
34 Radel et al., "Emotions and Gendered Experiences of Livelihood Migration."
35 Rikani et al., "More People Too Poor to Move."
36 Romanello et al., "The 2022 Report of the Lancet Countdown on Health and Climate Change."
37 Mora, "Over Half of Known Human Pathogenic Diseases Can Be Aggravated by Climate Change"; Oladejo et al., "Climate Change in Kazakhstan"; Intergovernmental Panel On Climate Change (Ipcc), *Climate Change 2022—Impacts, Adaptation and Vulnerability*.
38 Hall and Crosby, "Climate Change Impacts on Health in Remote Indigenous Communities in Australia."
39 Fritze et al., "Hope, Despair and Transformation."
40 Gebhardt et al., "Scoping Review of Climate Change and Mental Health in Germany—Direct and Indirect Impacts, Vulnerable Groups, Resilience Factors"; Hall and Crosby, "Climate Change Impacts on Health in Remote Indigenous Communities in Australia"; Berry, Bowen, and Kjellstrom, "Climate Change and Mental Health."
41 Vercammen et al., "Eco-Anxiety and the Influence of Climate Change on Future Planning Is Greater for Young US Residents with Direct Exposure to Climate Impacts"; Pinchoff et al., "Coping with Climate Change."
42 Cox et al., "Doses of Neighborhood Nature."
43 Pearce and Parncutt, "Quantifying Global Greenhouse Gas Emissions in Human Deaths to Guide Energy Policy."
44 Knez et al., "Before and after a Natural Disaster."
45 Hansen et al., "Global Warming in the Pipeline."
46 Rising et al., "The Missing Risks of Climate Change."
47 Stengers, *In Catastrophic Times*. P. 15.
48 Gignoux and Menéndez, "Benefit in the Wake of Disaster."
49 Kemp et al., "Climate Endgame."
50 Dehingia et al., "Climate and Gender."
51 Cuartas et al., "Climate Change Is a Threat Multiplier for Violence against Children."
52 Uenal et al., "Climate Change Threats Increase Modern Racism as a Function of Social Dominance Orientation and Ingroup Identification."
53 Stephenson, "Against Climate Barbarism: A Conversation with Naomi Klein."
54 Blumenfeld, "Climate Barbarism."
55 Stephenson, "Against Climate Barbarism: A Conversation with Naomi Klein."

3 Can We Do This? Embarking on Transformational Social Change

Despair sucks. I hate it! That feeling of complete and utter helplessness that feels like hands reaching up out of the pit and pulling me into the dark, sticky, suffocating mud. When I am feeling despair, it festers in the pit of my stomach, taking away my appetite for food, for hope, for any sensory input really. Despair is that voice in my head that says 'I am no David, and this beast is so much bigger and uglier than Goliath.' Despair is a self-fulfilling prophecy, as with so many other destructive stories we tell ourselves. Despair tells us we can do nothing about this thing we care about, whatever that thing is, and so we do nothing. Despair is just that, however, a story, and we can rewrite that story. When I feel despair creeping in, I recall the many Dene, Cree, and Anishinabe peoples, some of whom I have had the honour to meet, who share the horrors they have experienced, from residential schooling, to police violence, to missing and murdered loved ones, and who nonetheless wake up to greet the sun, prepare food for their families, come to ceremony, and resist colonialism, every single day. And so I get up, greet the sun, feed my family, and seek to rewrite the stories I tell myself.

Despair looms when we see no option, when we conclude that nothing can be done. There are for sure many manifestations of the climate emergency that we no longer have the ability to prevent. The ice sheets of Antarctica are already beginning to collapse, unleashing a rise in sea levels over the course of the next century that is irreversible. Millions, perhaps billions, of marine organisms have died due to warming oceans, and migrating salmon have been literally boiled alive in their spawning grounds. Homes, entire villages, have burnt to the ground. And yet, despite these and so many other irreversible outcomes that have already begun to unfold, there is so much more that remains that is deserving of love and protection, not least of which are the meaningful lives of yet-to-be-born humans and non-humans. Rewriting our collective story, however, requires transformational social change, and the monumental scale of this project invites despair. In this chapter, my goal is to convince you that, while mourning for our losses is absolutely warranted, despair is not, not yet anyway. I do so by contemplating *how* system change on such a monumental scale can occur.

Can we turn this proverbial ship around, and limit global warming to some threshold of human liveability—1.5 degrees Celsius seems to be behind us now, but every increment of new warming avoided is millions of lives saved—while adapting to the impacts already emerging? There have been all too many unhelpful responses to this question, some in the 'definitely not' category, while others are of the 'well obviously' variety. The 'definitely not' stories certainly do not inspire efforts to do anything but wait, as just discussed, but the 'well obviously' stories have exactly the same effect: let's just wait for technological ingenuity to wave its magic wand. The most helpful, and realistic, responses fall somewhere in between, in the 'maybe' zone. This is the zone that motivates action: efforts to identify and confront the specific mechanisms that lead us towards no, and to identify and upscale those that can lead us towards yes. Maybe is discomforting though, in part because maybe implies uncertainty, which is always unsettling, but also for the very reason that it provokes a response, a response that begins with unsettling questions about our beliefs, our lifestyles, our morality, ourselves. Well, get used to it my friend; discomfort may not feel good, but moments of discomfort are moments of learning,[1] so get uncomfortable, and know you are not alone.

My position falls firmly into this maybe zone. Based on current understandings of social change processes, we do have the ability to turn this ship around, *theoretically*. Have we done it before? Not exactly: we have no historical precedent for the speed and scale at which change must now occur if we are to avoid a warming-induced collapse of liveability on earth. Alas, addressing global warming is not simply a matter of excising CO_2 from our social systems, leaving them otherwise unscathed, nor is it a matter of putting Dr. Evil in jail (although I do have a few candidate jailbirds). Eliminating

40 *Can We Do This?*

greenhouse gases implies changing everything, as Naomi Klein[2] puts it. Global warming will absolutely change us; it already has, and we explored some of those processes in the previous chapter. Will we change global warming? By which I mean, will we undergo the *transformational* social changes required to curtail the social practices that generate greenhouse gas emissions and destroy the planet's carbon sinks? That remains to be seen, but as I argue throughout this book, one of the most important, and least studied, mechanisms governing social change is human emotionality.

Before we dive into emotionality, however, let's draw on some other knowledge sources in order to confirm for ourselves that yes, indeed, transformational social change happens, and sometimes those changes emerge in response to crises of various types. *How* crisis-induced change occurs, by what mechanisms, and the degree to which those mechanisms can be leveraged, however, are more difficult questions, in part because, at least in my own discipline of sociology, we have until recently not bothered to ask.

The Building Blocks of Dynamic Systems: Structure and Agency

Let's start with a clarification of the two most important elements of any system that we need to grapple with in order to understand system change: structure and agency. All systems can be described in terms of structure and agency, and there are some intriguing parallels across all types of system in this regard. Here we will focus specifically on social systems, albeit with full awareness that social systems are deeply entangled with ecological systems (if they were not, we would not be in our current mess), and the caveat that discussions about structure and agency across the social sciences are muddled, to say the least.

Structures are difficult to study, even difficult to talk about, because we can't see them, at least not directly. We only know of their existence by virtue of their effects. I like to think of structures as patterns of resource distribution—including material resources, information, and social capital—governed by an implicit set of rules, some would say myths, that have emerged over time, perhaps from small beginnings, but have come to be so deeply embedded in our social systems that many of us aren't even aware of the fact that we are abiding by them. Because social structures are overlapping, both with each other and with ecological structures, dynamics in one structure can affect others. Challenges to patriarchy, for example—a structure particularly important to our responses to the climate emergency that I will discuss in more detail in Chapter 5—can spill over into gendered divisions of labour, which in turn imposes shifts within capitalism—another particularly important structure discussed in Chapter 5. Structures are not easily changed, although change they do, not only as a result of these inter-structural interactions but also as a result of internal dynamics.

Institutions represent a system layer below structures. They are often described as entities that persist despite the movement into and out of them by individuals. Institutions are more readily identifiable, with names (the U.S. Supreme Court), identifiable office holders (the judges and their staff), and discrete boundaries (the U.S. border). Institutions also constitute discrete clusters of identities, norms, and practices, and an ethos—encompassing a unique set of moral strictures or guiding beliefs—around which members orient themselves, which provide a sense of stability, routine, and familiarity.[3] By extension, the withdrawal of support for an institution's identity, norms, practices, or ethos is a critical threat to institutional stability—the U.S. Supreme Court has experienced a heavy dose of scrutiny in recent years, for good reasons.

Another way to think of the difference between structures and institutions is that institutions are at least hypothetically escapable. I can remove myself from academia. I can choose to grow my own food and thus remove myself from the agri-business industry. There are certainly some important exceptions here: incarcerated persons do not have the ability to remove themselves from the prison-industrial complex. But structures are inescapable; we are all de facto members. There is no means of hiding from structures, of creating bubbles in which they are absent, as much as we might try. Even one's ability to initiate a feminist collective is 'allowed' or not by the patriarchal structure within which such initiatives are imagined.

Why do we build institutions anyway? According to the historical record, note anthropologists David Graeber and David Wengrow,[4] institutions are an ancient feature of human social life, and have contributed to the coevolution of genes and human behaviour. Those institutions, in effect, establish the norms and rules that allow those social structures to persist. Humans are the only species on the planet that creates institutions. Drawing from Peter Richerson and Robert Boyd,[5] our unique capacity to create, embrace, and belong to institutions can be traced back to the importance of group belonging for human survival. In other words, our innate desire to belong serves as the motivational force enabling institutions to persist. What holds a specific institution together is the emotional commitment to that institution of a set of adherents.[6] By interacting with other institutional members who embrace the same ethos, the same set of norms and practices, we receive positive emotional rewards in the form of a sense of self-validation, ontological security, and solidarity, which further reinforces a personal commitment to that institution.[7] Most of my academic colleagues hold the production and sharing of knowledge in high regard, and have ascribed to a set of norms, like researcher autonomy and methodological transparency, and practices—peer reviewing, thesis defences, participating in the conference circuit, and so on—based on the belief that these norms and practices are necessary for the production and sharing of knowledge. The financial supporters of those institutions, including states and private donors, and the individuals willing to pay tuition to pursue a degree, must also buy into this narrative, and holding a university degree remains a badge of honour, and a pathway to employment. Institutions in turn shape us. By prescribing identities, roles, beliefs, and practices, institutions inform our sense of self and shape the meanings we form of the information and events to which we are exposed.[8] According to Maxim Voronov and Klaus Weber,[9] one's "sense of self is conditioned and sustained through participation in social practices that are institutionally appropriate for their roles" in that institution. We each, as such, look to institutions to answer the question *Who am I?*

That allegiance is never completely stable, however. Our enormous capacity for reflexivity[10] means that one's commitment to a given institution can be retracted, and if such retractions spread, the institution becomes unstable. Therefore, personal commitments to those identities, roles, beliefs, and practices must be actively reinforced. Daily engagement within institutions includes systems of sanction and reward that serve this purpose. As with allegiance, these also operate primarily through emotionality. As we will discuss in Chapter 4, pride, shame, and guilt ensure compliance while rendering deviance enormously costly. Institutional rituals are another means of encouraging the continued revitalization of emotional commitments. Rituals generate shared emotional energy—recall the last time you watched a parade, viewed your national team in the Olympics, or attended the after-work barbecue—and in doing so, they generate a sense of common identity and shared understandings of reality among groups, some of which are too large to allow for direct social interaction,[11] as large as nation-states, ethnic diasporas, and religions. These processes are discussed further in the section below.

Structures operate at a higher level than institutions, and residing below this layer are organizations. An organization is considered more ephemeral than institutions, and more likely to be identified by the specific individuals who are members of that organization. If I start a business, a local bookshop, say, that business would be an organization. Both institutions and organizations are shaped by structures, indeed institutions and organizations do the work of operationalizing the myths that empower those structures. If there are no entities in society taking up and embodying the rules and practices prescribed by a given structure, then that structure cannot be said to exist. Structures are activity-dependent; they have influence only when we interact with them.[12]

A given social system, or society, is thus constituted by a historically and regionally contingent constellation of interconnected structures, institutions, and organizations. The highest rates of dynamism typically happen at the organizational layer, the lowest rates at the structural layer, and institutions fall somewhere in between. Representative democracy, for example, is a structure that is empowered by the myth that my rights and concerns can be incorporated into government decision-making by an elected representative. This prescribes, among other things, the formation of political parties. The U.S. Democratic Party, which has been around for about a hundred years, is an institution within which this myth is embodied. The Biden-Harris Political Action Committee, on the other hand, is an organization, with a relatively short lifespan, and attached to specific, identifiable individuals. The formation and activities of organizations can either reinforce or challenge those institutions, which in turn reinforce or challenge structures. As readers have likely already concluded, these distinctions are quite fuzzy! I agree, and debates regarding the dividing line between structures and institutions, or between institutions and organizations, will no doubt continue. But I find this three-layered depiction of social systems quite useful nonetheless for contemplating system change.

As it happens, depictions of agency are just as muddled. Sociological treatments of agency tend to converge on one key tenet: to be an agent is to have an effect, in some way, in the world around us. Convergence stops here, however. Most but certainly not all sociologists would add two additional criteria. First, to be an agent is to act *with intent*, which limits the capacity for agency to conscious, sentient beings. Second, agency presupposes that such actions are taken autonomously, rather than by force. According to these criteria, does an organization have agency? Many organizations have an effect, and organizations generally have stated intentions. On this basis, some say yes, an organization has agency. Others say no, however, because an organization is not a sentient being, it is the members of that organization who develop those stated intentions and act on behalf of that organization. Does a rock have agency? A virus? Western scholars by and large would say no, but there are notable exceptions, including Bruno Latour and his influential actor-network theory that includes pride of place for non-human actants, as well as many feminist and Indigenous scholars who wholeheartedly reject the arrogant notion that we humans are the only agents in this lively universe we inhabit. Frankly, the COVID 19 pandemic provides compelling evidence for the latter group.

I find these debates interesting and absolutely meaningful. However, for my purposes in this book, which is focused on the potential for transformational social change, I wish to focus attention on the agency of humans, whose agency manifests as autonomous, deliberate actions undertaken in the pursuit of an actor's unique set of interests and values, for the purpose of either defending something of value under threat, like an old growth forest, or manifesting a desired result, like a sharing economy. This is not to say that other entities do not have agency, and some of those entities have important roles to play in climate-society relations. Nor does it deny that we humans are embedded in relations with non-humans and with the planet. However, this book is about we humans taking better care of and responsibility for our relationships with nonhuman beings and our planetary home, which demands transformational change in our social systems. The current poor state of these relations is entirely on us, namely those of us who have been complicit in a Western project of domination for personal gain. And, as stated by Rom Harré,[13] 'if one's aim is really the transformation of society, the place to act is at the point where the people actually generate the roles and acts that are constitutive of institutions and other realities.' Social change happens when agents change the rules. One thing we can agree on, however, a truth that becomes integral to confronting the climate emergency and will be remarked upon throughout this book: the capacity for agency varies tremendously, by individual, by social identity, and by context.

Social Change, from Incremental to Transformational

As with all dynamic, complex systems, societies are regularly subject to forces of change. Most of those changes are *incremental*—ports become silted, new cultural schemas surface, a species goes extinct, a new zoning law is enacted, rivers change course, businesses fail. Most of these, furthermore, occur at scales below the system level. In other words, the outcomes of those changes tend to be contained, to a political jurisdiction perhaps, a watershed, or an economic sector. Incremental changes may nonetheless culminate in transformational change, if given enough time. Incremental responses to global warming in the form of shifting consumer preferences might, if given enough time, culminate in a low-carbon transition. Perhaps, if a package of incremental strategies intended to reduce emissions and support adaptation were initiated in 1980, or 20 billion CO_2 tons ago, those strategies may have been enough to avoid a climate emergency. As has become glaringly clear, however, given the state of the climate emergency as of 2024, after decades of inaction, incremental responses will be too little, and too late. Therefore, we are interested here in evaluating the prospects for *transformational* social change.

Transformational change is broader in its effects than incremental change. By transformational change, I am referring to the comprehensive re-writing of materialities, rules, beliefs, and practices associated with one social order, and their replacement with a new set of materialities, rules, beliefs, and practices, such that no member of that order is untouched. So, a household—a type of micro-social order—can undergo transformational change with the unfortunate death of one of its members; an ecosystem can undergo transformational change with a significant shift in precipitation regimes, or the extinction of a sentinel species; a scientific discipline may experience transformational change when its defining paradigms evolve. Given the global scale of the climate emergency, and the equally global scale of the socio-ecological systems with which it is entangled, we are talking about transformational change in all structures, institutions, and organizations that support practices that produce emissions or compromise carbon sinks, at the global scale.

Transformational change also implies irreversible change. Disruptions, which may be quite substantial interruptions to routines, might be the impetus for transformational change, but they may also be temporary, followed by a recovery phase that involves rebuilding institutions in pretty much the same fashion as before the disruption. The COVID-19 pandemic was a substantial disruption to societies everywhere, but whether or not that disruption transpires into transformational change at the institutional or structural level, or whether recovery efforts are geared towards a return to some pre-pandemic 'normal,' is an open question.

All complex systems are susceptible to transformational change—forests become savannas, mobile hunter-gatherer communities become sedentary farming communities, dictatorships become democracies, private automobile-based transportation systems replace public transit-based systems. Many historians refer to the Renaissance as a particularly noteworthy moment of transformational social change for Europeans and their descendants. The Renaissance was responsible for shifting Europe from the Middle Ages to Modernity, marked by the elevation of science and the state over the Church, and the replacement of feudalism with capitalism. For a more recent example, the changes wrought upon modern social systems by the internet continue to unfold, but I think it is fair to say that those changes have been transformational, given its effect on how information is shared, business is conducted, and communities are formed.

Each of these examples of transformational change has unique features, with respect to their emergent outcomes, and also with respect to their points of origin, their triggers. These moments of historical transformational change have been initiated through political, technological, cultural, and environmental prompts. The internet was clearly unleashed by a technological trigger; technology also had a strong role to play in the emergence of agriculture, and the private automobile. Most transformational changes are brought about by a confluence of prompts, however. A new technological innovation, after all, is not likely to become entrenched unless the cultural and political ground has been prepared to allow it to take root. Some instances of transformational social change, like the collapse of the Roman Empire, likely involved all four.

Those prompts can be further characterized in terms of whether they emerge externally, or are internal to the system in question, albeit those distinctions work better in our conceptual frameworks than they do in empirical research. Is the climate emergency an externally imposed crisis? Certainly, for some, such as a rural farming community in Mexico, or Arctic sea ice, but the climate emergency unquestionably has been created by social practices, and therefore it is very much internal to our global social systems. There is an additional differentiation, arguably even more important to this discussion: some instances of transformational change could be said to be *autonomic*—in other words changes that are unintended outcomes of system dynamics. The internet is once again a useful example: While it most certainly had its identifiable creators—a few computer scientists in the U.S., the U.K., and France trying to figure out a way to get computers to talk to each other—for all intents and purposes, it basically happened *to* the rest of us, fomenting dramatic changes that were never the intention of its creators, much like Frankenstein's monster. Rom Harré[14] refers to such moments as accidents of history—the unplanned coincidence of critical elements causing systemic change.

Other changes are *deliberate*, the outcomes of planned, intentional, goal-orientated pursuits. Revolutions represent deliberate, intentional, collective pursuits of social change according to a specific set of goals. This differentiation between accidental and intentional change is similarly problematic in its operationalization as the differentiation made between internal and external origins of change prompts, partly because the outcomes of such purposeful projects almost never go entirely according to plan. To the contrary, as is to be expected of complex systems, observed changes are always an emergent outcome of the interaction of multiple dynamic elements. In the words of Margaret Archer,[15] whose works I will return to below, 'society is an unintended consequence.'

The profusion of unintended consequences does not mean that intended outcomes never come to pass, however. The American Revolution achieved the goals of independence as envisioned by revolutionaries, along with a host of consequences that were not according to the script. The Renaissance absolutely inserted a new appreciation for art in European cities, as intended. It also unleashed the 'Age of Discovery,' a wonderfully euphemistic term for the colonization of Turtle Island, along with so many other Indigenous homelands across the globe—not necessarily the intention of the Medici's, and other Renaissance founders.

This is what we are ultimately shooting for: Deliberate, intentional, transformational change. A particular, and particularly rare, form of social change. And it needs to happen at the global scale, and lightning fast! The Renaissance unfolded over the course of three centuries; we have maybe three decades. Lofty? Yes, unquestionably. Impossible? We won't know until we try, but we do have a rich record of research that can give us some clues, and guidance.

Dispelling Unhelpful Assumptions

So, big changes are not unheard of. But *how* does transformational social change happen? Sadly, many social sciences, including my discipline of sociology, have had the unfortunate inclination to answer this question with 'it just does.' Even early scholars of social movements and revolutions tended to presume that such collective actions were a more or less autonomic response to social conditions. Or environmental conditions. In Gerhart Lenski's grandiose Ecological Evolutionary Theory, he argued that social transformation is driven by pressures of the environment that in turn force the selection of new technologies. Another variation of this response is Progressivism, a presumed linear process of social evolution, towards continuous betterment: technological advances, economic growth, increases in longevity are presumed, not explained.

Although Progressivism has thankfully fallen out of favour, particularly given its historic racist and colonial undertones, its traces remain evident today, particularly in the form of technological optimism. And across the social sciences, with some notable exceptions, there also remains a prevailing tendency to presume a spontaneous, deterministic link between disruption and social change. Ulrich Beck,[16] for example, speculated that direct, personal experience with extreme events like floods and fires would awaken an 'involuntary enlightenment'—such events would serve as a wake-up call, in other words, an opening for re-consideration of current social patterns and practices. Direct experience with such disasters certainly can be a wake-up call of sorts. Some recent studies have indicated that direct experience with extreme events can elevate climate concern—but mainly among those who were concerned already, and even responses that should be seen as elementary, such as re-zoning to accommodate rising storm surges, are fraught with conflict. We see plenty of examples of determinist thinking across climate dialogues too, with action recipes aplenty that include ingredient lists, but no instructions as to how to go about baking the bread, as if simply knowing that, for example, diets with a minimum of beef and dairy have lower carbon footprints will lead everyone to change their food habits. Many but not all of us are just as likely to clamour for a return to business as usual after disaster as we are to be enlightened, and just as likely to rationalize our continued beef and dairy consumption despite our full knowledge of its impact on the climate.

Another word for this set of assumptions is functionalism: the presumption that the behaviour of a social entity, whether an institution or an individual, can be explained (determined) by its effectiveness. So, an individual, organization, or institution will always behave in ways that are in their best interests. If I learn that smoking is bad, I will quit smoking because I am a perfectly rational actor; if the Ministry of Environment learns that a certain policy does more to harm the environment than improve it, then surely the policy will be redacted. To be sure, the "functionality" of a given social entity is an important consideration in analysis across the social sciences. New health information can and does motivate behaviour change in some, and the declining effectiveness of that environment ministry can, under certain circumstances, lead to a withdrawal of trust and legitimacy among influential stakeholders, which can in turn weaken that institution to such a degree that the institution must change course, or risk collapse. But it would be folly to presume the inevitability of either the withdrawal of trust and legitimacy, or the responsiveness of that institution to such a withdrawal, any more than we can presume that the production of new climate science findings alone will lead to much of anything at all, except perhaps a bump in the H-factors of some lead authors.

For these reasons, there are plenty of examples we could turn to of maladaptive structures, institutions, organizations, and personal practices. The capacity of some institutions to avoid legitimacy crises through information management, the relative power of stakeholders, and the capacity of that institution to undergo transformational change are three of the most important variables influencing such outcomes, but there are others. Likewise, there are no limits to my own capacity to engage in behaviours that are not in my best interest. Current anthropological accounts of institutions under threat point to the decreasing likelihood of abandoning a course of collective action the higher the historical investments in that course of action in terms of economic and social capital, increasing the potential for failure to adapt to changing circumstances.[17] The fact that decades of society-threatening environmental and climate disruption have gone largely unattended despite our full knowledge of those disruptions is blatant evidence of the capacity for institutions to be non-responsive to crisis and instead charge ahead, right off the cliff.

In other words, the possibility for failure, for collapse of institutions, and also communities, ecosystems, and entire societies, is real. Indeed, social institutions are so resistant to change that some researchers have concluded that they don't. Instead, it may well be that in order for social change to occur, existing institutions must disintegrate and new institutions emerge to replace them, particularly when the underlying ethos of those institutions is no longer compatible with the new demands being placed upon them. What institutions are destabilizing—losing their functionality—as a result of climate change? Nothing much really, just those involved in energy, agriculture and food, transportation, housing, urban planning, health, emergency response, environmental protection, finance, immigration, and foreign policy. Is an oil and gas corporation capable of reinventing itself into a renewable energy corporation? Can the World Bank become a bastion of degrowth? Or do these institutions need to dissolve and be replaced? While it is safe to say that the scale of global warming-induced change that has already begun to unfold is likely unprecedented in human history, the possible agentic responses to those destabilizing forces nonetheless encompass multiple potential scenarios, from orchestrated transition to ecologically and climate-friendly systems on one end of the spectrum, to catastrophic loss of life and social breakdown on the other.

Some More Constructive Theories of Change

Thankfully, not *all* social scientists resort to simplistic accounts of social change, and useful work is being done across the academy. There are many places to look, but here I highlight a few specific threads, being spun by scholars working with the historical record, with resilience theory, and with reflexivity theory, particularly the work of Margaret Archer.[18]

Insights from History

While historical accounts of societal collapse due to shifting ecological conditions—self-inflicted or otherwise—are several, the historical record also includes indications that many societies have also instituted changes in response to ecological and material crises. Jared Diamond's *Collapse*[19] is well-known, and valuable despite his exclusive focus on social transformation failures, because of the elements he identifies as collapse facilitators, including, for example, increasing investments in support for an elite class, or just plain cultural rigidity—a failure of belief systems to respond to changing circumstances. There are many less well-known works that are equally informative; however, some of which do indeed point to instances of successful transformative change, although I will mention just a few.

In a rare, macro-study drawing on an extensive historical database of social responses to crisis, Daniel Hoyer and colleagues[20] clarify that environmental and climatological crises are by no means a death knell for societies; in fact, many societies have endured where others have collapsed. The authors identify a few crucial ingredients for resilience, including equality, which presumes the provision of social welfare to those in need, and also social cohesion, which tends to go hand in hand with equality. Such a finding certainly does not inspire confidence in the resilience of many social systems today!

Most historical studies are case-specific, allowing us to go into greater depth. For example, in an extensive study of the settlements in the American Southwest over the past 2,000 years, Rebecca Dean[21] shows that settlements responded successfully to environmental variability through migration and resettlement, and the gradual replacement of hunter-gatherer systems with subsistence agriculture, both of which coincided with increases in sociocultural complexity. Dean attributes these instances of adaptation to the flexibility of knowledge systems: the retention of social memories of past events, combined with the uptake and sharing of new information pertaining to environmental conditions, distilled into *schema* that serve as a codified and transferable unit of cultural knowledge. Developed further by Roderick McIntosh, Joseph Tainter, and Susan McIntosh,[22] the dynamic evolution of cultural schemas is a central organizing principle explaining societal responses to climatic change, which entails 'innovation in the form of experimental recycling or reinvention of curated knowledge of past climate experience and of economic and sociopolitical strategies that previously provided solutions,' which become 'coded in the foundation legends, beliefs, and material cues that serve in turn to structure a society's perception of its environment.'

The evolution of cultural knowledge systems is marked by shifts in the legitimacy of cultural schemas. Cultural schemas, as with all institutional norms and practices, are only effective to the extent that they are perceived as legitimate, which gives them their authority. That legitimacy is maintained in part through their continued utility in explaining current events. They run the risk of losing their legitimacy otherwise. Shifts in cultural knowledge systems, however, involve not just the loss of legitimacy of incumbent schemas, but also the availability of alternatives. Adaptation thus depends on the rapid uptake of new information and circumstances into social memory. From McIntosh, Tainter, and McIntosh[23] again:

> The potential adaptive strength of complex societies lies in the fact that such societies create more variations on cultural schemata to filter data and to guide the entry of knowledge into social memory. To generate diversity, complex societies need to encourage (or at least tolerate) subgroups' knowledge collection so that pooled information will encompass a greater range of climate variation and store a larger set of past behaviors.

Social structures that constrain the generation and uptake of new information (or spread misinformation) would be considered more vulnerable. Which is exactly where we appear to be today. As David Graeber and David Wengrow[24] lament, we find ourselves 'stuck' today, incapable of imagining alternative social arrangements to the ones we have inherited, despite the fact that our history is replete with such alternatives, including stateless societies, and egalitarian societies. Indeed, as the Davids illustrate in their vastly ambitious *New History of Humanity,* 'the earliest known evidence of human social life resembles a carnival parade of political forms, far more than it does the drab abstractions of evolutionary theory.'[25]

To be sure, there are several missing ingredients in these accounts, such as consideration for who is engaged in the production and sharing of knowledge, who has the power to challenge or defend the legitimacy of those problematic cultural schemas, and introduce new ones—power dynamics that help to explain why we are stuck today. Nonetheless, the means by which our knowledge systems adjust to changing circumstances—in other words, our collective capacity for learning—remains an important element in these stories of social change.

Resilience Theory

Social systems are not the only systems that undergo change, and ecologists also bring important insights. Resilience theory, rooted in ecology and borrowing heavily from complexity theory, has inspired a new genre of research into ecosystems, notably including the human inhabitants of those systems. The theory is rooted in the understanding that nature is complex, evolutionary, and adaptive, exhibiting both chaos and order, continuous and discontinuous elements, and abrupt, episodic change, describing a complex landscape that is highly dynamic yet capable of self-organization.[26] This intermingling of chaos and order is captured in the Panarchy framework of ecological system change, first introduced by Lance Gunderson and C.S. Holling in 2002,[27] although Holling had been developing the ideas that were to culminate in this co-authored work for decades prior. The Panarchy framework depicts ecosystems in a continuous and dynamic state of flux, which marks an important point of departure from earlier conceptions of ecosystems as tending towards a climax state defined by balance and stability. According to the Panarchy framework, the vitality of a given system cannot be defined as the ability to reach a single climax end-state at all, but rather in terms of adaptation processes, amounting to a continuous evolution through four functional phases: exploitation, conservation, release, and reorganization, leading to the next phase of exploitation.

That reorganization phase, however, marked by the breach of certain tipping points, can either precipitate collapse, in which case the old system is replaced by a new system, or that reorganization can assume the form of 'bouncing back' to its original character. This 'bouncing back' potential is what is referred to as resilience. Within these systems, resilience theorists differentiate systems in a manner very similar to my discussion of social structures, institutions, and organizations above, with micro-, meso-, and macro-level scales ordered hierarchically, on the basis of the level of dynamic change featured in each, with faster dynamics found at the micro-level, and slower dynamics featuring at higher scales. The lower scales provide feedstock for change, while the higher scales provide the system with a semblance of stability. These subsystems nonetheless affect each other through cross-scale feedbacks. Because the micro-scale is the most dynamic, resilience scholars postulate that one particularly important route to system change is via the upscaling of innovative and novel changes that emerge from within niches at the micro-scale, which then spill over, through the meso- and macro-scales. The character of those dynamic processes shapes the likelihood for collapse or resilience. Gunderson and his colleagues identify three necessary qualities to support the resilience of a given ecological unit: (1) the system must accumulate resources (rather than deplete them) over time; (2) it must contain both destabilizing forces, which are necessary for maintaining diversity, resilience, and opportunity, as well as stabilizing forces for maintaining productivity and biogeochemical cycles; and (3) evolutionary processes must be sufficiently dynamic and accommodating of diversity to generate novelty (like those cultural schema mentioned earlier).

Resilience theory is attractive in many respects, particularly in its articulation of system dynamics, and the conditions placed upon the divergent outcomes of resilience or collapse. So attractive, in fact, it has been taken up—one could say bastardized—in public administration and social planning, with mixed results. According to this theory, a resilient society would be described as one for which each scale is relatively autonomous—allowed to operate at its own pace—but the system is both protected from above and invigorated from below through information feedback mechanisms. In rigid systems controlled by the narrow interests of an elite minority, change can only occur when a triggering event unlocks the social and political gridlock of the higher levels. Resilience scholars have struggled to give credence to agency, however, and students are then left with a response to the question 'how does change happen?' that still amounts to some version of 'it just does.'

Reflexivity Theory

I am attracted to reflexivity theory because of its sophisticated treatment of agency, and by extension of structures, and system change. One of the reasons my own discipline has such a hard time explaining social change—ironic isn't it?—has to do with our difficulty in articulating the inter-relations between structure and agency, with those scholars who are focused on macro-social change tending to pay far more attention to structures than to agency, and vice versa. One of the more influential sociologists in the past 50 years, Pierre Bourdieu, is a case in point: he offers a very helpful means to articulate the enormous influence of social structures on everyday lives and practices, in the form of what he refers to as *habitus*.[28] He leaves questions regarding how social actors negotiate with social structures, and how those structures change as a result of those negotiations, to other scholars to work out.

In order to give agency its due in our theoretical understanding of social change, I have been particularly persuaded by the scholarship of the late Margaret Archer, whose work falls within critical realism. Archer departs from many of her sociological contemporaries like Pierre Bourdieu by giving credit to our large brains, which provide us with the cognitive capacity to differentiate our selves from our circumstances; to exercise reflexivity, the 'internal conversations' we continuously engage in as a means of navigating our lives. So yes, we each are born into socio-cultural contexts—a habitus if you will—in which a given set of social structures and institutions feature, and those social structures can become so embedded in our daily lives that we are not always aware of their influence. But we are not simply automatons responding to structural prompts. Indeed, we are continuously deliberating, often with ourselves, sometimes with others in our network: Should I buy those pants? How should I respond to that email? Why didn't I get selected for that job? Why did Russia invade Ukraine? What would compel a person to say goodbye to homes and loved ones and climb aboard an over-crowded boat about to cross the Mediterranean, perhaps never to return? Through these deliberations, we also imagine for ourselves a set of projects, based on our unique sets of personal identities, predispositions, values, and life histories—whether that project is to become a professional athlete, pursue a low-carbon lifestyle, decolonize our schools, or find a life partner. And we each have the capacity for awareness, for deliberation, and confrontation with those social structures, as those structures simultaneously enable and constrain the pursuit of our projects. Our deliberations may help us determine the best route to the achievement of our goals, or we may foresee so many constraints in our path that we decide to give up before we begin. The influence of structures, then, manifests in their causal power to enable, and to obstruct, our projects.

This is not a one-way street, however. According to Archer, individuals also have causal powers to affect structures: one's reflexivity may encourage that agent's continued endorsement of those social structures, as when a woman determines to abide by prescriptions for gendered social roles. Or it may motivate attempts to strategically navigate those social structures in pursuit of personal projects, as when an underprivileged youth pursues every source of financial aid available to go to college. Or, that reflexivity may lead one to problematize those structures, and pursue projects intended to change them, as when a family embarks upon a buy-nothing month, or a climate-anxious youth engages in protests intended to give fossil fuel corporations their due. And sometimes—certainly in the last example—those individuals determine that the most effective means of realizing that ambitious goal is to do so collectively. One important (but insufficient) trigger that enables actors to move from unreflexive institutional reproduction to the more reflexive orientations of problematization and action for change is encountering contradictions between beliefs and lived experiences—that is, situations that both require a reflexive distance from established patterns and enable greater imagination and conscious choice.

This is where the social change rubber really hits the road. Social structural change begins with individual-level problematization of the existing institutional order or some feature of it, and the imagination of projects the intention of which is to challenge that institutional order. When the number of individuals committing to such challenges by taking individual actions grows, the aggregate effect of those projects can lead to social change, as when, for example, a growing number of consumers ditch animal products, and grocery stores, restaurants, and food processors initiate changes in response. The speed and scale of change are elevated substantially, however, when those individuals seek to pursue projects collectively.

The potential for purposeful transformational social change is thus initiated at the point of social interaction, in which grievances are aired and shared. By engaging in social interactions, we each are changed—perhaps our perspectives are broadened, we feel validated, empowered—and simultaneously our contributions to those interactions change others. Social change happens when these social interactions amount to the ideation and validation of challenges to incumbent beliefs and practices; those challenges can then spread through social networks, taking the form of direct challenges to the legitimacy and performance of institutions, and by extension the structures they embody.

This is by no means an inevitable outcome, or even a typical one. The fact of the matter is, we all tend to find a great deal of comfort and ease in our habits and routines. Deliberation takes effort, after all. Sharing our deliberations with others can make us feel vulnerable, and we don't necessarily have the stomach for it—we'll come back to this issue in the next chapters. Even oppressive situations can become a sort of comfort, of security, the security that comes with knowing what to expect. In other words, most of us don't do change well. For such a historically adaptive species, our resistance to change has a certain irony about it. We dislike change so much in fact, that we convince ourselves that change doesn't happen, and are repeatedly surprised when it does. We don't give any thought to a prenuptial agreement in the throes of love, we don't prepare for disasters when the weather is fine, we convince ourselves that today will be much like yesterday, and tomorrow much like today. Until it isn't, and an illness, a lost job, a broken refrigerator, or wildfire, runs headlong into our expectations.

But, it is those less typical moments, when change is in our sights, that make history, not the more typical ones. For Archer, moments of social structural change, what she refers to as structural morphogenesis, are defined by the positive feedbacks that result when individual agents are mobilized into new groups, particularly when they amount to a re-grouping of social relations, as when settler and Indigenous knowledge-keepers come together, or labour unions and environmental organizations. Those re-groupings of mobilized agents can then produce feedbacks that enable the re-distribution of material resources, and the introduction of new ideas as cultural resources, that in turn motivate increases in reflexivity among others, who then engage in their own social interactions, and so on, and so on. In order words, those cultural schemas mentioned earlier, which appear to be so important to social change, do not surface of their own accord. You and I must first dream them up, and then give them room to move.

Rolling with Maybe

What can we take away? One, change does happen *to* us, all the time. Global warming is happening to us in a big way, right now—simply avoiding global warming's effects and keeping things as they are is not an option available to any of us, although some have much more insulation than others to be sure. But if we want to actually steer, rather than drift, then we need to embark upon a course of purposeful transformational social change. Second, my answer to the question 'can we do this?' is a strong maybe. Although we have no direct historical corollaries, our current state of knowledge regarding social change supports the *possibility* if not the *probability* for transformational social change. We don't know for sure whether efforts to pursue a future in which runaway climate change is contained will be successful, but we do know for sure that such a future will not happen if we don't try. Third, there are many mechanisms that are invoked in social change processes, but two in particular are ideas and agency. Ideas, or cultural schemas if you prefer, are the tools, but the tools need to be crafted and wielded by agents, working collectively.

What drives human agents? Reflexivity theory tells us 'how' people come to engage in their projects/practices, and acknowledges our individuality, the source of imagination, and innovation. But we need to delve into emotions to help us learn 'why' different people reflect and respond the way they do to their circumstances. Emotions can enable or disable project development, and can influence the shape of, and success of, efforts at collective action. In Jonathan Turner's words, 'emotions are the energy that sustains or changes social reality.'[29] Our emotions not only define our reflexive processing of the world around us, our attachment to our projects, and the social relations within which such projects are pursued; they are in fact our greatest resource, particularly those emotions that govern cooperation. In short, the most important piece of this social change puzzle, not covered in this chapter but taking centre stage in the rest of this story, beginning with Chapter 4, is our emotionality.

Notes

1 Wilson, "Discomfort."
2 Klein, *This Changes Everything*.
3 Voronov and Weber, "The Heart of Institutions."
4 Graeber and Wengrow, *The Dawn of Everything*.
5 Richerson and Boyd, *Not by Genes Alone: How Culture Transformed Human Evolution*.
6 Voronov and Vince, "Integrating Emotions into the Analysis of Institutional Work."
7 Voronov and Vince; Creed et al., "Swimming in a Sea of Shame."
8 Creed et al.
9 Voronov and Weber, "The Heart of Institutions." P. 458.
10 Voronov and Weber.
11 Turner and Stets, "Sociological Theories of Human Emotions."
12 Archer, *Realist Social Theory*.
13 Harré, "Social Reality and the Myth of Social Structure." P. 119.
14 Harré.
15 Archer, *Realist Social Theory*. P. 165.
16 Beck, "Living in the World Risk Society."
17 Janssen, Kohler, and Scheffer, "Sunk-Cost Effects and Vulnerability to Collapse in Ancient Societies."
18 Archer, *Realist Social Theory*; Archer, *Being Human*; Archer, *Structure, Agency, and the Internal Conversation*.
19 Diamond, *Collapse*.
20 Hoyer et al., "Navigating Polycrisis."
21 Dean, "Social Change and Hunting during the Pueblo III to Pueblo IV Transition, East-Central Arizona."

22 McIntosh, Tainter, and McIntosh, *The Way the Wind Blows*. PP. 24–26.
23 McIntosh, Tainter, and McIntosh. P. 28.
24 Graeber and Wengrow, *The Dawn of Everything*.
25 Graeber and Wengrow. P. 119.
26 Gunderson and Holling, *Panarchy*.
27 Gunderson and Holling.
28 Bourdieu, *Outline of a Theory of Practice*.
29 Turner, *Human Emotions*. P. 177.

4 What Are Emotions and Why Should We Care?

I Feel Therefore I Am

> The soul breathes through the body.
>
> Antonio Damásio[1], 1994 p. xxi

Across the social sciences, researchers agree that we humans are complex and emotional beings, and of course, we have all had an opportunity to confirm this for ourselves based on personal experience. Yet many of those same researchers are inclined to resort to some version of a rational-utilitarian model of human behaviour to explain everything from the decision to participate in a social movement to having children, and this is certainly true

DOI: 10.4324/9781003380900-4

of the climate social sciences. As depicted by Helena Flam,[2] the 'rational man [sic] of classical economics is desirous, calculating, consistent and selfish.' Emotions, not to mention identities, positionalities, life experiences, and agency are all conveniently boxed up and stored in a dark corner of our offices and laboratories and labelled 'Do Not Open.' The appeal of rational-utilitarian models of human behaviour should be obvious: they are just so clean and easy to work with! If we presume that people are fully informed and make decisions by methodically tallying up the personal costs and benefits of different courses of action, then we can safely ignore many messy features of human behaviours as well as social structures. The rich complexity and diversity of humanity is smoothed, blended, washed away for the convenience of simpler models. The researcher can simply presume (1) the interests of the actor—usually money, power, and happiness—and presume (2) that the individual's (fully informed and calculated) actions will be in pursuit of these interests. End of story.

Thank goodness humans do not in fact operate in this way, because we would likely be extinct by now, having no faculties whatsoever to address collective action problems. The consequences of such a one-dimensional model of human behaviour resonate far beyond academia. As will be discussed in the next chapter, such individualized, instrumental characterizations of what makes us each tick are also enormously, and not accidentally, compatible with capitalism. Indeed, a capitalist economic order presupposes atomized, predictable, rule-following, self-interested, utility-maximizing agents. Such models *prescribe* as much as they presuppose, to such an extent that emotionality is broadly treated as a disruptive force that should be constrained. Unless, of course, those emotions favour competition, consumption, and other behaviours that endorse the current economic order.

The other problem is that such a model of human behaviour is woefully inadequate. Considering the enormous body of evidence refuting it, this model should be put to bed once and for all. While we certainly engage in cost-benefit analyses on a regular basis (is the price of those new pants really worth it? Can I afford organic produce?), our deliberations are by no means strictly utilitarian, and the 'costs' and 'benefits' include non-personal preferences as well as personal ones. Our reliance on a rational-utilitarian model of human behaviour leads not only to limited explanatory power in scientific research; it also leads to ineffective policies, when consumers and citizens simply do not behave as predicted (prescribed). Attempts to control emotions for the sake of maintaining social order and efficient decision-making may well have the opposite effect some or even much of the time.

A rational-utilitarian model of human behaviour is partly rooted in highly erroneous dualistic models of reason and emotion, and by extension of society and nature, as if we can choose whether to think or feel, but never both at the same time. In short, to develop a richer, fuller understanding of human behaviour at the individual and collective levels, as they pertain to responses to the climate emergency but in all other fields of social inquiry too, we need

What Are Emotions and Why Should We Care? 59

to bring emotionality to the fore. In this chapter, we therefore take a deep, and deeply interdisciplinary, dive into the study of emotionality to explore the questions: *What are emotions? Why do we have them? Why do they matter?* Before we jump in, though, we need to acknowledge one very important caveat to the empirical research record reviewed here. The vast majority of research in the affective sciences by and large relies upon empirical studies using population samples that describe a very limited slice of the global social milieu, namely Western, educated, industrialized, rich, democratic (WEIRD) samples,[3] and the very small number of studies that depart from this trend pose some serious challenges to the generalizations that have been drawn from this research record.[4] The application of an intersectionality lens—forefronting distinct social positionalities defined by race, ethnicity, class, gender, age, and so on—to the study of social responses to the climate emergency is absolutely essential, and therefore we must also apply an intersectionality lens to our reading of this research record as well.

What Are Emotions? Contending Definitions

After having delved into emotions research for some years now, I can offer my full agreement with Pfister and Böhm's observation that the concepts of emotion and emotionality are anything but concrete, agreed-upon terms. To the contrary, 'emotion is a word used in the vernacular to refer to loosely related phenomena. The conceptual confusion to be observed in the literature about definitions of emotion manifests that fact.'[5] The key point of difference among the multiple definitions is where they fall along a spectrum from determinist to constructivist, positions which mostly but not entirely correlate with specific disciplines. Each perspective offers a not inaccurate yet partial view, warranting an interdisciplinary lens.

Towards the determinist end of the spectrum, we have, for example, neuroscientific depictions of emotions as strictly genetically inherited biological functions of the nervous system, which humans have acquired to serve as a biophysical valuation system developed to quickly assess opportunities or threats, both within the body (hunger) and the environment (danger). For these researchers, it is the biophysical realm of emotions that matters most.[6] As George Stefano[7] states, 'emotion provides the motivation for action, the mechanism to limit reason in a timely survival related manner and a coping strategy for dealing with other humans and animals while simultaneously modulating involuntary physiological functions in an appropriate manner.' Rolls[8] adopts slightly more instrumental language, without departing from the main tenets laid down by Stefano, LeDoux, and others, describing emotions as responses to reward and punishment. The important points offered by this perspective include recognition that emotions are indelibly embodied, biophysical phenomena, and are for the most part autonomic—while we can certainly choose how we behave in response to our emotions, we do not *choose* to be sad, happy, or angry. As well, while some forms of genetic variation, as well as physical and mental harm, may compromise the emotionality of some individuals, and as we will discuss further below, life experiences certainly generate variations in our emotionality, humans share 99.6% of their genetic heritage, and therefore our affective systems are also shared.

The late Dutch psychologist Nico Frijda takes us a little further afield from this end of the spectrum, interpreting emotions as a much broader valuation mechanism, constituting what has come to be known as appraisal theory. For Frijda,[9] emotions serve as our means of assessing the personal relevance, or meaningfulness, of information or experiences in our environment, on the basis of the cultural norms and values we have been socialized to embrace, and personal life experiences. In other words, emotional reactions are never solely focused on the objects or phenomena towards which our emotional attention is directed, but also to their subjective meaning for the observer, allowing for multiple, unique individual responses to the same phenomenon, such as being disciplined by the boss, or reading a news story about climate change. Appraisal theory thus broadens our purview of the determinants of the relevance of stimuli to which we are exposed beyond survival and reproduction, and an acknowledgement of the culturally and socially constructed nature of each of those determinants: for example, group belonging and our status within that group; and a sensitivity to social justice.

Sociologists who focus on emotions tend to land on the constructivist end of the spectrum, focusing on the social-structural and social-interactional bases of emotionality, premised upon the understanding that emotions emerge and evolve in the context of social interactions.[10] From a constructionist view, emotions are bodily experiences whose expression cannot be separated from socio-cultural contexts.[11] Edith Stein, whose dissertation was published in 1916, has been recognized for the foundational work she has done to bring emotions into our sociological imagination. She centres emotionality upon a continuous dialogue between individual predispositions and communal experiences.

A more fruitful way of enhancing our understanding of what emotions *are* is to focus on what they *do*. The short answer is, well, everything. Emotions play an integral role in all forms of decision-making and behaviour. Pfister and Böhm[12] group different emotions into four categories based on what they do, including those which: inform pleasure and pain for preference construction; enable rapid decisions while under time pressure; allow for focused attention on complex problems; and motivate commitment concerning morally and socially significant—in other words hard—decisions. More succinctly, Rimé[13] (p. 4) states, 'emotions are mechanisms that make us learn something.' Our expansive and complex human emotional palette can be attributed to our expansive and complex human brains. The neurosciences have provided extensive contributions to our understanding of the intricacies of emotion centres in the human brain, illustrating the multiple and distinct neural sources of different emotions which—and many may find this surprising—constitute *the most dominant* physiological processes in the brain.[14]

Interesting as it is, I will not be diving too deeply into this work. Instead, we can focus attention on two primary regions of the brain that are most centrally involved in the generation of emotions. On the one hand, among the oldest (evolutionarily speaking) parts of our brain is the limbic system. Colloquially known as our 'hind brain,' or 'reptilian brain,' this includes the amygdala and hippocampus, along with other components. The limbic system generally does its thing continuously and subconsciously, until it activates alertness to potential dangers or opportunities, in which case those signals are conveyed to other parts of our brain and thereby into our consciousness. Clearly, our limbic system is critically important to survival. Without it, I doubt I would have survived my teen years. But it can also be very tiresome, inducing the generation of lust and excitement, or fear, anxiety, distrust, and aggression, even in situations where such reactions are unhelpful to the situation at hand. Its detection of danger, for example, is not always particularly accurate (consider our propensity to be afraid of the dark, or new foods), which means the emotional responses triggered by our limbic systems can be out of alignment with any actual threats, and may be an inherited reaction that was far more relevant at other points in human history (darkness means lurking predators we can't see, unfamiliar foods may be poisonous), or a remembered threat from the past (like a car accident when you were a child), rather than present circumstances. An over-active limbic system may be detrimental to relationships, information sharing, and cooperation, not to mention mental health. This is so because the limbic system is connected to everything else, including those parts of the brain we use to retain memories and make decisions.

The other part of the brain I want to highlight is, in contrast, the newest part of our brain, the cortex. The cortex includes temporal, parietal, and occipital lobes as well as the frontal cortex and pre-frontal cortex. In addition to giving us our prominent foreheads, the components of the cortex are the origin of many of our uniquely human attributes (at least most neuroscientists, such as Panksepp,[15] think they are unique to humans, but research on other species may eventually prove us wrong). In effect, the cortex provides us with the basis for self-awareness and reflexivity—what some scholars refer to as Mind—and in turn the ability to reflect upon our own mental states, and the mental states of others. The capacities offered by the cortex include language, memory, deliberation and planning, and the *regulation* of those emotions triggered by our limbic system, so we do not necessarily act on our first impulses. The cortex is also the source of emotions that are uniquely developed in humans (earlier-noted qualifier still applies), such as pride, shame, and empathy, all of which will be discussed in detail further below. Tiresome in a different way than the limbic system, the exercise of our cortex consumes relatively much more energy than other parts of our brain. If you find yourself snacking more when you are trying to write a term paper for class, blame your calorie-hungry cortex. For this reason, hard-wired as we are to conserve energy—another inherited survival instinct that does not always serve us well—we are inclined to avoid making use of this part of our brain, despite its critical role in complex thinking. Those individuals who have experienced or continue to experience trauma and distress, or are simply too pre-occupied with their crazy-busy lives, have a particularly low mental energy reserve and are thus more inclined to respond behaviourally to signals from our limbic system.

A number of researchers focused on decision-making refer to these two brain regions in terms of a 'dual-process' model, which Joseph LeDoux refers to as the 'high road' and the 'low road,' and Daniel Kahneman calls System I and System II, depicting a 'slow-thinking' route (via the cortex), and a 'fast-thinking' (limbic) route to decision-making.[16] Work by Kahneman and others draws attention to the fact that despite the enormous deliberative capacities afforded by our cortex, we don't make use of them as often as would be ideal, and thus our limbic system is more often engaged, not only in those situations for which it is perfectly well-suited to guide rapid decision-making, like jumping out of the way of an oncoming car, but also in those situations for which it is not, like discussing global warming over Thanksgiving dinner. Importantly, the full employment of those frontal cortex regions of the brain is needed to support decision-making under conditions of uncertainty and complexity, and also to override initial affective reactions that induce unfavourable social behaviours (Eek that brown person with all the earrings and tattoos is scary! I'm going to grab my purse and move to another part of the bus).

All too often, particularly in contemporary Western societies, we just do not have the energy to make the investment of effort that the cortex demands. This can lead to poor decision-making of the types laid out, for example, in the classic decision heuristics typology developed by Tversky and Kahneman.[17] Notably, focusing solely on individuals, as is often done in the affective sciences, overlooks the relevance of social-structural contexts that impose conditions upon individuals that *produce* trauma, distress, over-work, and over-stimulation. In one recent study, Muscatell[18] showed that the limbic systems of individuals of lower socio-economic backgrounds are frequently on overdrive, in response to their devalued and disempowered status. Many other current features of Western social systems also discourage deliberation and reflexivity in other, perhaps more subtle ways, particularly the general tendency towards top-down decision-making in many of our institutions. With simple rules to follow, whether to be a good employee, a good student, a good Christian, or a good environmentalist, no further contemplation is required.

Dual process models of decision-making can be useful, but ultimately, they are simplifications of neural processes, and have received their share of critique,[19] in part for the implied but very erroneous separation of emotional and cognitive brain regions.[20] All of the brain's affective regions have essential and distinct roles to play in day-to-day decision-making and behaviour, by drawing our attention to matters of concern, and then giving us the capacity to deliberate upon those matters of concern in order to guide decision-making and action, putting to bed previous notions that reason and emotion work at cross purposes, or are even separable beyond the level of a conceptual model. Antonio Damásio has developed what I find to be a more useful way to think about emotions, what he refers to as the somatic marker hypothesis: emotions enable us to 'mark' for cognitive attention certain stimuli (events, information, situations) from the panoply of daily stimuli with which we are confronted, and assist us with evaluating the potential consequences of different courses of action. In other words, 'emotional information serves as a relational map of our environment.'[21] In many ways, intelligence is tantamount to emotional intelligence. What does intelligence entail after all? Acuity in the storage and organization of memories, invoking those memories and new information in the practice of inductive and deductive thinking, and making and sticking with decisions that are the result of careful reasoning. All of these mental activities involve emotions. Without the emotional capacity to hone in and prioritize, to commit to a course of action, our ability to make decisions from 'do I run or fight?' to 'what's for dinner?' to 'should I join a social movement?' would be sorely lacking.

Why Do We Have Them? The Evolutionary Origins of Human Emotionality

We can't fully grasp our emotionality, or respond to the question 'why do we have emotions?', without talking about where they came from. I am compelled to begin this section with a bit of foot stomping. When I mention the name Charles Darwin in my graduate seminars, many students visibly squirm, expecting me to next display 'scientific' images of male and female brains perhaps, or move into a lecture on why poverty is natural. Likewise, I have many colleagues in sociology who wince when I mention evolution, a term which for many in my home discipline recalls images of eugenics, essentialism, and all sorts of creative efforts by actors in the academy and beyond to justify assertions of privilege over others. Those images—which include some particularly dark moments in our history—should not be forgotten. Indeed, one of the key lessons provided by this history is particularly relevant to our social responses to the climate emergency: the potential for manipulation of scientific information for deleterious political ends.

On the other hand, I find the inclination to hold the scientific knowledge regarding evolution itself accountable for this history, rather than the (primarily Western, white, male) actors who put their partial interpretations of that knowledge into such deleterious use to be perplexing to say the least. Critiquing early treatments of social evolution for their functionalist undertones and vehemently rejecting racist sociobiology are absolutely warranted. However, denying that humans are a species, and like all other species subject to and shaped by natural selection, has been terribly costly to many fields in the social sciences and humanities, providing fuel for human exceptionalist tendencies—the tendency to view humans as separable and superior to the rest of nature—and thus hampering inquiries into social-ecological systems. The persistence of human exceptionalism in the social sciences and particularly my own home discipline of sociology has distanced many social scientists from consideration of our embeddedness within nature, including human evolution, as well as other sources of biological and ecological knowledge that are directly pertinent to who we are, what we do, and where we are headed. To be sure, biological determinism has no place in the social sciences; but neither does social determinism. Our genes, our biology, our relations with land, water, and nonhuman species all matter integrally to the phenomena that social scientists study, but this is not to say that they *determine* human behaviour and social change.

Foot-stomping over, let's get started. The drivers of biological evolution may be difficult to observe, and their expression becomes ever more convoluted in complex societies, but they are straightforward in principle: as with every other organism on the planet, humans pass on genes to the next generation, at least those of us who survive to reproductive age and then do in fact reproduce. That replication process tends to generate genetic variants, which then generate new traits, some but not all of which are beneficial to survival and are thus naturally selected for over (a very, VERY, long) time. Evolution is demarcated by adaptation to changing environmental conditions—the ultimate driver of that selection process. No adaptation, no survival.

How did we come to embody such a complex and on occasion burdensome suitcase full of emotions? To make a very long story woefully short, we developed an expansive emotional palette, in large part because we humans are generally lacking in many of the physical features that aid in the survival of other animals, like fast feet, poisoned fangs, and sharp claws. Our best bet, if we were (are) to survive, was (is) to form groups and coordinate our efforts, and our ability to do so is enabled by our emotionality.[22] Cooperation in turn supports knowledge-sharing: individuals living in groups are not dependent solely upon their individual capacity to acquire knowledge about the environment first-hand; they have at their disposal a wealth of knowledge, tools, and skills shared by others. Group belonging, and the learning and sharing it enables through cooperation, became an instrumental adaptation strategy for our human ancestors. And survive we did: Despite some rather timid beginnings 200,000–300,000 years ago (accounts differ on the timeline), *Homo sapiens sapiens* would appear to have been spectacularly successful at survival, evidenced by our occupation of nearly every possible habitat the planet has to offer, and our monopolization of the planet's resources, much to the detriment of other species. Indeed, all indications are that we have been *too* successful and are rapidly compromising the carrying capacity of the earth to support us, but let's park that thought for the moment.

With a focus on groups and networks, the story of human evolution becomes a story of culture as much as it is a story about genes. If there is one thing that human evolution scholars (mostly) agree on, it is that the relationship between genes and behaviour is two-way, and the vehicle that travels back and forth is culture. Work by scholars such as evolutionary biologist Peter Richerson and anthropologist Robert Boyd,[23] as well as sociologist Jonathan Turner,[24] has brought the role of social groups and hence culture into evolutionary view. Pagel[25] refers to culture as our species-level 'operating system,' and the groups and networks that store and share knowledge, ideas, and practices are survival vehicles that are just as important as our physical wellbeing. As a result, social selection has become a much stronger and more rapid driver of human evolution and social change than our genes. In fact, research has shown that natural selection has responded directly to social environments: the larger the social group, the larger the neocortex.[26] Kim Hill and colleagues[27] suggest a potential explanation: increased network size among our ancestors, detected archaeologically by the emergence of long-distance flows of tools and raw materials, likely led to greater exposure to novel ideas worth learning.

Throughout human evolution, natural and social selection thus worked in tandem to favour the acquisition of human capacities for survival, which includes foremost capacities for creating and sustaining social bonds—cooperation requires a commitment to others after all. But how do we humans manage to work together to hunt, gather, protect ourselves from predators, and rear families without killing each other? This is where emotions come in, including in particular empathy, as well as pride, shame, and guilt, but there are many others.[28] Rather than the dual-process model described in the previous section, I prefer to think of the affective circuits in our brain as sedimentary layers, deposited over human evolutionary history, with each new layer not replacing but being built atop older layers, resulting in what at times becomes a hot mess of emotional baggage, but also a fantastic toolbox to support not just individual survival, but also societal survival, by supporting collective action. So, on the one hand, as summarized by Panksepp,[29] our brains are wired to seek out positive personal rewards, which can lead to excessive materialism and greed, as well as aggression towards those who appear to threaten our wealth, our liberties, our freedom, or even our comfort zones. This can manifest as xenophobia, and antagonism towards groups that do not appear to share our values and beliefs. Our desire for autonomy, furthermore, can lead to aspirations for social dominance, leading to the types of power politics with which we have become all too familiar. On the other hand, we are also capable of love, compassion, sharing, and empathy, supporting acts of generosity that are enormously costly to ourselves.

An important lesson from this body of scholarship is what it tells us about what it means to be human. Evolutionary history does not determine our fates. This is the case, ironically, because of our evolutionarily developed capacity for Mind. We nonetheless carry the imprint of our evolutionary history, in the form of several core propensities that help to explain our feelings, thoughts, and actions—the understanding of which only enriches our capacity to deliberate and change our courses of action. For one thing, we tend to gravitate towards positive emotions and avoid negative emotions. In other words, we are motivated to experience things like enthusiasm, love, self-worth, efficacy, and confidence, and to avoid feelings like sadness, anxiety, shame, and guilt.[30] By extension, Turner[31] and others have improved our understanding of some basic prerequisites for mental wellbeing, a more fulsome depiction of what human thriving entails that takes us way beyond the oft resorted to material indicators, like dollars and calories. These include the need to belong and closely related to this, the need to have our personal sense of self, our identity, validated by our peers. It includes the need for expressiveness, for reflexive autonomy or the ability to establish one's goals independently, as well as efficacy—the belief that one has the capacity to pursue those goals, that one's actions matter. I would add here an additional element not often acknowledged in the social sciences until recently: our integral connections with the natural world, and the personal costs we pay when we are alienated from those connections.

Social systems that allow for such needs to flourish may just be more stable than those which do not. These realizations lead us to another important set of questions: how do current social conditions support these basic needs for human thriving, or not? In a controversial yet compelling re-examination of the archaeological record, David Graeber and David Wengrow[32] describe these basic needs in the form of three freedoms, the freedom to move, the freedom to disobey, and the freedom to form social relationships. Their relative absence in modern society has, for Graeber and Wengrow, limited our capacity to pursue deliberate social change, leaving us 'stuck' with maladaptive rules and practices. This premise becomes particularly important as we explore the dialectical relation between emotionality and institutions in the next chapter. But first, let's take a closer look at some main players in our emotional palette that come into play in our responses to the climate emergency.

Some of the Main Players in Our Emotionality Palette

Human emotionality describes a lot of territory, some of which is more easily labelled than others. I may be acutely aware that something is bothering me, but am I feeling anger, rage, disgust, a bit of all three, or none of the above? Pinpointing that feeling can become an exercise in splitting hairs that for most of us is just not that important. To a certain extent, the labels we assign to emotions are constructs deployed for the sake of communication, not least of which among researchers, but those constructs nonetheless do have descriptive value. I will make no attempt to cover the hair-splitting gamut of human emotions here, but rather narrow in on a selection that appears to have particular bearing on personal and collective responses to the climate emergency: guilt, shame, pride, and empathy. Each of these emotions is particularly important to enabling or constraining individual and collective behaviour. They are thus of particular interest for the prospect of social change.

Guilt, Shame, and Pride

Guilt and shame are both negative emotions that we would all rather avoid feeling altogether. Why do we even have this particular set of emotions? Because, even though we have developed strong inclinations to support things like altruism, we still have our egos, and desire for autonomy. That's not all bad; that same ego is the source of things like innovation and rebellion. But if we allow those egoistic traits to rule, all forms of collective action would be compromised.

Guilt and shame are important counterforces to selfishness and aggression, of which we are all capable, and have their useful purposes for survival, but are deleterious to family and group formation[33]. To the extent that even the prospect of feeling shame or guilt can prevent certain behaviours—in other words anticipated emotions, rather than experienced emotions—they can be an enormously effective means of ensuring social conformity and hence group cohesion.[34] With these emotions in play, we don't even have to go to the trouble of meting out punishment for bad behaviour. In certain social contexts, they can induce in us a motivation to behave in ways that can be highly consequential for ecological and climate wellbeing too, by supporting pro-environmental behaviours. But they can also do the opposite, motivating the overconsumption of material goods, including the purchase of identity-linked goods like clothing, and lavish gift-giving. Ever show up at a dinner party empty handed, only to discover all other guests have a bottle of wine or some other gift for the host in hand? You'll only do it once, I'm sure. Or perhaps you will feel compelled to eat meat even though you really want to be vegetarian, because your grandfather is at the barbeque this evening. Or the anticipation of shame might prevent you from speaking out among peers, even if you deeply disagree with where the conversation is going.

The distinctions between guilt and shame are commonly missed, but they are important. While guilt is action-specific, directed at a specific act of harm, shame is an evaluation of the whole self, of one's self-worth.[35] Guilt motivates avoidance of harm upon others, and serves as a check on our selfish or aggressive inclinations,[36] favouring cooperation and fairness. Guilt is guided by social context, but is governed more so by one's internal moral compass. That moral compass also guides other emotions—particularly anger and rage when we observe violations of our moral compass by others, but it is particularly pertinent to guilt. I feel guilty when my actions conflict with what I believe is right, as when I drive to the market rather than taking my bike. In other words, guilt is to a certain extent individualized—some of us but not others feel guilty about stealing or having an affair, even while occupying the same culture. Few people feel guilty about driving the car to the market, or about the animals killed for dinner, but some most certainly do. Guilt also is group-specific. In highly unequal societies, we may be less inclined to feel guilty about harming those above or below us, either in terms of class, race, ethnicity, or gender, than we would about harming others whom we deem our equals. Stealing the stapler from work might actually feel good, a way of 'sticking it to the company'; stealing from your neighbour not so much.

While both guilt and shame are negative emotions, they operate quite differently. Shame, along with pride, is an indicator of social competence—conforming to a social group's normative expectations, or failing to do so[37]—and by extension they are personal triggers that indicate our success in establishing social belonging. As such, these emotions are entirely governed by social relations in specific group contexts, and not about committing an act understood to be unacceptable. It's getting the answer right when called upon in class, or getting it wrong; getting compliments on your new outfit versus realizing you are underdressed for an event. Shame involves viewing oneself in another's eyes. Quoting Barbalet,[38] "with shame, self is necessarily qualified by the other." While shame and pride are governed by other's views of oneself, one's susceptibility to either is also shaped by one's personal sense of self. Shame indicates a discrepancy between one's self-perception and the feedback received by others, while pride indicates consistency between them.[39]

By enforcing conformity shame supports group cohesion, but it can also have deleterious social effects, for the shamed individuals, and also for society, as these individuals will seek to disengage from shame-inducing social situations. Shame is a great silencer, an effective way to smother one's agency. The problem is, for some groups those shame-inducing social situations include school, work, church, and public spaces, particularly in social settings characterized by power dynamics. Sociologist Theodore Kemper[40] has suggested that shame is especially likely to arise in relations in which the individual inclined to shame holds a lower status than others. For this reason, shame can be weaponized by those with power to great effect, and with horrible consequences, taking the form of bullying on the schoolyard, doxing on

Twitter (sorry, I mean X), or racial profiling by real estate agents, teachers, and police officers. Shame is also not solely situational—many of us can find ourselves in a rather long-lasting state of shame, particularly if we are objectionable in some way to the dominant social group in ways that we cannot rectify—based on our poverty, skin colour, sexual orientation, religion, or ethnicity. Traumatic experiences, like sexual abuse, can also invoke a long-term state of shame, even in relations with others who remain unaware of the historic acts themselves.

Given how awful they feel, and how much we are averse to negative emotions, relying solely on negative emotions like shame and guilt to support group cohesion wouldn't be particularly effective—why belong to a group that only makes you feel bad? Pride is the counterweight, an important positive emotion that also supports group belonging and cooperation. Pride is about positive evaluations by others, received when we do things recognized by others as worthy, and in accordance with cultural norms. We experience pride when our actions are well-received by others, or more generally, when we are made to feel validated. In a social system in which norms support cooperation, those actions can include giving, and other forms of personal sacrifice in service to the community. Humans are capable of extraordinary acts of altruism, perhaps in large part due to the role of pride. Heroism in many cultures is defined by sacrifice: think going to war, entering a burning building to save someone's life, or giving away all the meat from a successful hunt. These acts not only feel good, they enhance your reputation in the community. That in turn enhances your own prospects, for survival, finding a mate, or promotion. What's the best thing you can say at someone's funeral? 'He would have given you the shirt off his back!'

Like shame and guilt, pride can also lead to negative personal outcomes because it too is closely linked to culturally prescribed norms. So, big boys don't cry, women must be perfect mothers and never age, teenagers engage in risky behaviours to impress peers. Actions considered worthy of reward can become highly distorted, particularly in siloed cultural contexts. Like attacking the U.S. Capitol to impress your commander in chief. Or on Wall Street, where generating wealth at all costs, including tax evasion or theft, becomes a badge of honour.

Empathy

Empathy is just as important to social belonging and cooperation as are shame, guilt and pride; I would argue it is in fact the most important, which is why this section will be quite a bit longer than the previous one. Empathy is also a bit more complicated than shame, guilt and pride, which I will get into shortly, but in many ways, empathy can be understood as the ability to read the emotions of another, the ability to walk in another's shoes, or as stated by Clark[41] (p. 34) 'an imaginative leap into the minds of others.' For

Jonathan Turner[42] (p. 104), 'the ability to read emotions, anticipate how they will affect another's behaviors, and then make adjustments to these behaviors is what allows humans to cooperate,' manifesting in what Edith Stein called a 'we-mode' of being.[43] Empathy allows for the existence not only of families, communities, and societies, but also collective intentionality, in other words, collective problem-solving.[44]

Empathy has been a source of attention by sociologists dating back at least a century, and among philosophers for far longer. Empathy (Einfühlung) was a centrepiece of Simmel's *The Problems in the Philosophy of History*, for example.[45] Less known but equally germane is the work of German philosopher Edith Stein. Stein established the link between empathy and intersubjectivity,[46] which has only been reinforced in research since. Empathy scholarship in the ensuing years, across multiple disciplines but particularly the neurosciences and psychology, has refined our conceptual understanding of empathy, and substantiated those concepts with empirical research.

Scholars since Stein have recognized the multi-dimensionality of empathy, generating varying yet strongly overlapping efforts to articulate empathy's distinct components. Joseph Decety[47] has perhaps put the greatest amount of effort into developing a typology of empathy, consisting of three elements: affective sharing, empathic concern, and perspective taking. Because each element involves different regions of the brain, and is invoked in situ, each element may manifest in individuals in varying combinations and intensities in different situations. Importantly, they do not necessarily emerge synchronously; either of the three might serve as an entry point towards empathic behaviours. The likelihood for empathy to serve as a motivator to act in a caring manner towards others, however, is greatly enhanced when all three are activated.

Affective sharing refers to an individual's often involuntary inclination to resonate with the emotional state of another.[48] Flinching, for example, when watching someone trip and fall in the crosswalk, is a form of affective sharing. Affective sharing does not necessarily transpire into caring for another, however, unless it coincides with *empathic concern*, which describes the inclination to value another's wellbeing and hence express concern for their emotional distress (or joy for their positive emotional state).[49] The third element, *perspective taking*, involving the conscious effort to conceive of what another individual is experiencing, is the most cognitively engaged component of empathy, and that component understood to be most unique to humans.

These propensities are rooted in a set of neural capacities enabling attachment, self-other differentiation, and executive function. Our innate desire to form social bonds begins with child-parent relationships. This universal human drive provides us with the felt need to connect and form relations with others, and also the cognitive ability to apprehend others' affective states.[50] Self-other differentiation, also referred to as intersubjectivity, is the capacity to recognize one's own subjectivity, and hence also intuit the subjectivity of others, such that we each can grasp the mental state of others and differentiate those mental states from our own.[51] Prefacing this capacity is the elemental, embodied human capacity to associate actions with their perceivable effects,[52] which gives way to the ability to *anticipate* the effects of an act, both for oneself and for others, and thus allowing for anticipated emotions, mentioned earlier.[53] In the words of Shepard,[54] we are tuned to resonate with observations in our environment that are deemed familiar and meaningful. Perspective taking would not emerge in the absence of self-awareness, and by extension, self-other differentiation.

Empathy is not likely to motivate prosocial behaviours without a high level of executive function—the ability to self-monitor and reflect on thoughts and actions, allowing for mental flexibility and emotion regulation[55]—which enables one to express empathy towards another without being overwhelmed by the negative emotions that may surface.[56] Without executive function, negative emotions can become overwhelming, and we might be inclined to turn away from the source of distress and thereby avoid the negative emotions that it stimulates, rather than take action.[57]

While these basic cognitive and emotional prerequisites for empathy are inherited, as multiple studies involving children as young as newborns confirms,[58] their manifestation is indelibly shaped by social context. The surfacing of empathy requires a situation, an encounter with others, human or otherwise. Opportunities for social interaction presented by living in groups, or in close relation with nonhuman kin, provide the context for empathy to develop, and thus while empathy can be invoked virtually, empathy is most readily invoked during direct interactions, which allow for eye contact, the rapid assessment of similarities, and of another's emotional state[59]; assessments that are far more likely to be accurate and increase the prospects for intersubjective understanding than when interaction is via a phone call, or online.

Observing another being in distress is a common trigger for empathy, but perception of unfair treatment or injustice imposed upon another, even upon those outside of our immediate social group, is also a strong trigger for empathy, because injustice is an experience with which most of us are familiar (although to vastly varying degrees),[60] but also because sensitivity to fairness appears to be another of those hardwired traits, as any parent of toddlers knows all too well. Our inherent sensitivity to fairness, then, is what allows for empathy, and ultimately the basis of morality.

Another conditional factor influencing the likelihood for empathy to be expressed is social identity. We are more likely to express empathy towards others who are perceived to be similar to us, either in appearance, in values and beliefs, or mutual membership in larger symbolic social structures like classes, nation-states, and religions. We are simply more likely to have direct interactions with people similar to us, creating opportunities for empathy to emerge; but perceived identity overlap, combined with our ability to communicate values, experiences, and ideas through language, also allows for empathy to be expressed—and groups to be formed—among similars even in the absence of direct interaction.[61] Those beings and entities with which we identify can also include the nonhuman world.[62] Research has shown that feelings of relatedness and love for nature are crucial to sustaining pro-environmental behaviours.[63] In contrast, worldviews supporting the separation of humans from nature provide cognitive justification for nature's exploitation and assertions of domination, discouraging empathy towards the nonhuman world.[64]

Empathy is attributed to a wide range of prosocial and self-sacrificing behaviours.[65] Engaging in prosocial acts can generate positive feelings for an empath, particularly when the performance of such acts is perceived to have an impact by reducing the distress of another[66]; doing so has been associated with the release of dopamine.[67] But engagement in empathic acts is by no means inevitable; several other factors can intervene in the pathways between empathic feelings and action. First, innate personal capacities for expressing empathy do vary. Genetics play a role, with twin studies confirming genetic variation in expressed levels of affective sharing and empathic concern, although interestingly, not perspective-taking, which appears to be entirely learned.[68] Other studies have identified a positive correlation between oxytocin levels and empathy, explaining in part consistent empirical findings of differences in empathy between males and females.[69] The social environment, particularly during adolescence, also features prominently. Individuals who were raised in environments lacking in social interaction and emotional bonding show lower levels of empathic capacities in adulthood.[70]

Second, empathy activation when either executive function or efficacy is lacking describes an unstable psychological state that could lead to denial, or withdrawal.[71] These two intervening factors can also interact, such that feelings of powerlessness to help someone in distress elevate one's own emotional arousal, which can further diminish an individual's executive function: their capacity to manage those emotions.[72] Third, the actions motivated by empathy are not inevitably pro-social. Empathy may, for example, support the sacrificing of the many to save one identifiable, present victim.[73] Empathy can also be the precursor to anti-social acts, like racially motivated acts of cruelty and violence,[74] when such acts are perceived to be for the protection or defence of one's social identity.[75] Empathy in essence abides by our constructed divisions of the social universe into in-groups and out-groups, the means by which we attach our identities to certain collectives and just as actively differentiate ourselves from others[76]: our 'empathy maps,' to adopt Arlie Hochschild's[77] term. We engage in rapid categorization of strangers into Us and Them, based on observational cues.[78] For sociologist Dawne Moon,[79] 'emotional investments in the definition of a category with which one identifies can lead to a protective posture, reinforcing metaphorical tent walls to keep out threats to a group's cohesion or, potentially, survival.'

However, while our need for identity verification and many other aspects of empathy are more or less hardwired, we have enormous flexibility in our means of delineating in-groups and out-groups, seemingly carving up our social world 'at the toss of a coin,'[80] and recognition of this potential is growing. After reviewing mountains of empirical evidence on the history of humanity, including evidence of geographically expansive social networks and mobility that challenge notions of small, close-knit, kin-based groups as the basis of social life, notions that have supported theories about tribalism and its links to empathy, Graeber and Wengrow[81] insist 'we all have the capacity to feel bound to people we will probably never meet.' This means that, while our in-group/out-group delineations can be the source of anti-social behaviours, this flexibility may also allow for the opposite, motivating the deliberate expansion of our empathy maps.[82] As Christakis[83] shows, fluidity of group membership allows for cooperation and altruism to emerge across group boundaries. The most important means by which these boundaries can shift, however, is through perspective-taking,[84] which demands energy. Attempting to take the perspective of my own father is hard enough, but deliberately seeking to overcome negative feelings towards members of out-groups and embracing their perspective is sufficiently effortful to become a formidable barrier.[85] Cultures dictate the values that serve as the basis of shared meanings, making it difficult for a person embedded in one culture to take the perspective of someone from another culture that may prescribe an entirely different set of values. In short, it requires a high degree of personal development and cultural evolution working in concert for individuals to empathize with strangers to the same degree as they do with family and friends. As Breggin[86] speculates, there was likely little to no advantage throughout much of human evolution to inhibit suspicion and violence towards outsiders, and therefore those capacities, which are most urgently needed today, are comparatively lacking.

This constraint on empathy map expansion is furthered by the fact that performances of empathy are conditioned by social structures, which locate people in social compartments defined by class, income, race, gender, and occupation, and thus determine the likelihood for social interactions to occur.[87] Social media appears to have only enhanced this compartmentalization.[88] While history includes moments of high levels of collective empathy,[89] there are indications that our contemporary socio-economic structures have fostered particularly low levels of empathy, in part due to the high degree of inequity we are experiencing today, with stark differences in privilege, not solely based on material wealth. Highly racialized, gendered, and class social divisions describe an important set of compartmental boundaries. But living under such conditions of inequity can impose barriers to empathy in other ways. Most notably, inequity compromises the efficacy and executive function (mental health) of marginalized groups through repeated confrontations with many forms of oppression. Social inequality is also linked to increases in affective (as opposed to ideological) polarization,[90] the prevalence of which in turn produces negative effects for democracy, such as reduced levels of engagement—and consequently declines in issue diversity—in politics.[91]

The means by which empathy is performed is thus orchestrated, one might even say choreographed, according to institutional norms. Hochschild's empathy maps are to a significant degree culturally prescribed and maintained.[92] Gendered social structures provide a particularly clear example, with girls and women in many cultures socialized to care for others, while boys and men are socialized to avoid expressions of emotion at all.[93] Class represents another social structure that reflects distinct social patterns, with research indicating that members of upper classes are less inclined to express empathy than members of lower classes, and are even more inclined to commit harms upon others, by cheating and stealing, for example[94]; traits that research suggests develop in response to the accumulation of wealth, rather than the other way around.[95] Political ideologies also operate in ways that correlate with empathic behaviour. The values associated with conservative and liberal political ideologies map onto empathy in different ways, attracting adherents at least in part on the basis of their empathic predispositions, with liberals consistently expressing higher levels of empathy towards others, while conservatives are more likely to adhere to a social dominance orientation, with negative implications for empathy.[96] Liberals are nonetheless just as inclined to express antagonism towards out-groups, particularly as affective political polarization has become increasingly prominent.[97]

Beyond the Individual

Although much of the research just summarized centres on the individual as the unit of analysis, it should be clear from this review that our emotionality has everything to do with our relationships. This is relevant far beyond the study of emotions: as much as we in the West like to talk about leadership, ingenuity, and other individualized attributes as the levers of change—as if Louis Pasteur, Ursula K. Le Guin, Abe Lincoln, and Louis Riel each managed to conjure their achievements and contributions to society from the depths of their own private contemplation caves—our capacity for agency, for social change, even for the motivation and creativity that supports such action, is to be found in our relations with others.[98] Even our sense of personhood, of our individuality, is formed through social interaction. Our ability to engage in relationships, in turn, is enabled by our ability to read the emotions and intentions of others.[99]

We do so with the aid of that most basic element of empathy—a capacity many have called emotional contagion or entrainment.[100] According to He and colleagues,[101] 'humans both consciously and unconsciously transmit emotional signals that are essential for fostering social bonds and for maintaining good interpersonal relationships.' This capacity has been attributed to 'mirror neurons.'[102] When mirror neurons fire, we have the same emotional reaction when observing something happening to someone else as we would if it were to happen to us. Rizzolatti and Sinigaglia[103] explain, 'the mirror neuron mechanism embodies that modality of understanding which, prior to any form of conceptual and linguistic mediation, gives substance to our experience of others." Our brain waves become synchronized through interaction with others.[104] Fowler and Christakis[105] conducted an interesting study in which they observed that the happiness of a given individual is closely correlated with the happiness of other people with up to three degrees of separation in one's social network.

While researchers generally agree that mirror neurons are most likely to be triggered in direct, face-to-face interactions, recent research has evidenced emotional contagion can also travel through digital domains like social media.[106] In other words, even *imagining* an interaction can have a similar effect on mirror neurons,[107] which of course has been the source of livelihood for many an artist, author, and movie director. Emotional contagion, in combination with our linguistic abilities, supports a preference for storytelling as a particularly resonant mode of communication. We tell stories in order to share an emotional experience, and extensive research has shown that emotionally laden narrative-style messages flow especially rapidly through social networks.[108] In fact, stories appear to more readily engage those parts of our brains that process information and store memories than information provided in other forms.[109]

Our emotion-sharing capacity is also the foundation of crowd behaviour, allowing for the cultivation of what Emile Durkheim[110] called 'collective effervescence,' supporting group belonging and solidarity among large numbers.[111] Hence the success of emotionally evocative public rituals such as campaign speeches, 4th of July parades, and street protests at cultivating that we-mode of being that Edith Stein referred to, and motivating commitments to collective goals. The more strongly people identify with a group to which they belong, the more strongly they respond emotionally on behalf of that group.[112] Emotional and cultural capital are created when group members laugh together, celebrate together, succeed together, and cry together. Emotional contagion is also highly likely to transpire among individuals who have shared histories, values, and language, and among people who have shared the same experience, whether a soccer game, a terrorist attack, or a hurricane[113]; dramatic events that can serve to quickly generate new in-groups, or realign existing ones.

Social interaction also works the other way—inducing the management or suppression of personal emotions that are in violation of norms embraced in a particular cultural setting[114]; norms to which most of us all too readily conform. Why do we conform? To protect our group belonging, of course.[115] As Robert Sapolsky[116] says, belonging is safety. But it is more than this. The cultivation of group belonging goes hand in hand with identity formation—that sense of peace and stability that comes with knowing oneself. As such, that feeling of shame when we experience any form of discipline or sanction is, at its deepest level, a destabilization of our sense of self. As a result, we are each willing to go to great lengths to bring our sense of self into alignment with the feedback we receive from others.[117] The stronger the sense of belonging an individual feels towards a group—the more that belonging to that group becomes essential to that individual's identity—the more strongly they respond emotionally on behalf of that group. Those emotions in turn motivate action, including a willingness to commit atrocities towards members of out-groups.[118] The construction of out-groups goes hand in hand with the construction of in-groups, and our highly developed inclinations to do so become evident in children as young as toddlers. Members of out-groups then become objects of fear, dismissal, even hatred, and disgust.

The concept of emotional contagion should be accepted with one important caveat. Recall our personal uniqueness and agency: despite our common genetic make-up, we are not all sewn from the same cloth, and then there's that huge cortex of ours that allows for the personal capacity to evaluate, to deliberate. Individuals respond differently when observing the same events or expressions of emotions in others.[119] The moderating effect of individuality generates unique dispositions to respond to the events we experience in ways that may diverge from others. And while deliberate nonconformity is always enormously emotionally costly for the reasons just described,[120] our agency allows for such deviance, and that deviance becomes the means by which the shifting of group boundaries, collective action, and social change are all possible. Why some individuals are willing or able to take a bold stand for change while others are willing to support the status quo is an enduring question for the social sciences, but we all nonetheless have the same neural wiring that enables us to choose either of those pathways.

Some Summary Points

Emotions enhance survival, and enable both the formation of large groups and the ability to cooperate within those groups. They are the basis of morality, of ethics, and in many ways they are what makes our lives meaningful. As James Jasper[121] has argued, not only are emotions 'part of our responses to events, but they also—in the form of deep affective attachments—shape the goals of our actions.' Our emotionality explains how our bodies, including our brains, tend to interpret and respond to our circumstances, and it is deeply integral to the decisions we make. In other words, how we respond to the climate emergency is governed to a very great extent by our emotions.

On the other hand, emotions do not *determine* our actions. Neither genes, nor emotions, rule human thought and behaviour. Well okay, *some*times they do, as when we commit acts when angry that we later regret. Many other times, however, we think before we act, as when we choose to act constructively despite our anger, or pursue certain goals despite our fear, like standing up to oppression. And it is true, *some* people seem to be better at thinking before acting than others. Nonetheless, while our emotional reactions may be 'hard-wired,' our behaviour isn't. This is because, for one thing, we humans are also indelibly social—we read cues from each other, and this guides our behaviour as well as our beliefs. Our seemingly personal decisions are never wholly independent, they are interdependent—what do I think others are doing? How are my actions contingent upon the actions of others? For another thing, despite our sociality, we are each reflexive agents. Sensory systems signal the brain, but the brain can also send signals to the sensory system. I might feel fear or disgust instantaneously in response to a situation—an invitation to try a dish of crickets, say—but my frontal cortex can override that behaviour stimulus. Emotional pathways can shift. Just as anger can be replaced by forgiveness, climate apathy can pivot towards empathy and action.

The problem is, this emotional override takes effort, and increased demands upon our mental energy blunt our deliberative and empathic abilities. Just making our way through our modern lives can be mentally exhausting. In many ways, our highly evolved brains still have a long way to go to catch up with the modern social structures that constitute central features in our socio-ecological environment, with direct implications for our ability to tackle modern problems.[122] Those social structures, and how they come into play in our emotionality, are the subject of the next chapter.

Notes

1 Damásio, *Descartes' Error: Emotion, Reason and the Human Brain.*
2 Flam, "Emotional 'Man': I. The Emotional 'man' and the Problem of Collective Action."
3 Gurven, "Broadening Horizons: Sample Diversity and Socioecological Theory Are Essential to the Future of Psychological Science."
4 Phoenix, *The Anger Gap: How Race Shapes Emotion in Politics*; Webster and Albertson, "Emotion and Politics: Noncognitive Psychological Biases in Public Opinion."
5 Pfister and Böhm, "The Multiplicity of Emotions: A Framework of Emotional Functions in Decision Making." P. 7.
6 LeDoux, *The Emotional Brain: The Mysterious Underpinnings of Emotional Life*; Immordino-Yang and Damasio, "We Feel, Therefore We Learn: The Relevance of Affective and Social Neuroscience to Education."
7 Stefano, "Cognition Regulated by Emotional Decision Making." P. 1.
8 Rolls, *Emotion and Decision Making Explained.*
9 Frijda, "The Evolutionary Emergence of What We Call 'Emotions.'"
10 Bericat, "The Sociology of Emotions: Four Decades of Progress."
11 Fox, "Emotions, Affects and the Production of Social Life"; Turner and Stets, "Sociological Theories of Human Emotions."
12 Pfister and Böhm, "The Multiplicity of Emotions: A Framework of Emotional Functions in Decision Making."
13 Rimé, "Emotions at the Service of Cultural Construction."
14 Stefano, "Cognition Regulated by Emotional Decision Making."
15 Panksepp, *Affective Neuroscience: The Foundations of Human and Animal Emotions.*
16 Kahneman, *Thinking, Fast and Slow*; LeDoux, *The Emotional Brain: The Mysterious Underpinnings of Emotional Life.*
17 Tversky and Kahneman, "Judgment under Uncertainty: Heuristics and Biases."
18 Muscatell, "Brains, Bodies, and Social Hierarchies."

78 *What Are Emotions and Why Should We Care?*

19 Pfister and Böhm, "The Multiplicity of Emotions: A Framework of Emotional Functions in Decision Making."
20 Damásio, *Descartes' Error: Emotion, Reason and the Human Brain*.
21 Summers-Effler, Van Ness, and Hausmann, "Peeking in the Black Box: Studying, Theorizing, and Representing the Micro-Foundations of Day-to-Day Interactions." P. 456.
22 Andrews and Davidson, "Cell-Gazing Into the Future: What Genes, Homo Heidelbergensis, and Punishment Tell Us About Our Adaptive Capacity"; Gould, *An Urchin in the Storm*; Richerson and Boyd, *Not by Genes Alone: How Culture Transformed Human Evolution*.
23 Richerson and Boyd, *Not by Genes Alone: How Culture Transformed Human Evolution*; Boyd and Richerson, *Culture and the Evolutionary Process*.
24 Turner, *On the Origins of Human Emotions*; Turner, *Human Institutions: A Theory of Societal Evolution*; Turner, *On Human Nature: The Biology and Sociology of What Made Us Human*.
25 Pagel, *Wired for Culture: Origins of the Human Social Mind*.
26 Sapolsky, *Behave: The Biology of Humans at Our Best and Worst*.
27 Hill et al., "Co-Residence Patterns in Hunter-Gatherer Societies Show Unique Human Social Structure."
28 Turner, *On the Origins of Human Emotions*.
29 Panksepp, *Affective Neuroscience: The Foundations of Human and Animal Emotions*.
30 Collins, *Interaction Ritual Chains*; Cozolino, *The Neuroscience of Human Relationships: Attachment and the Developing Social Brain*.
31 Turner, *On the Origins of Human Emotions*.
32 Graeber and Wengrow, *The Dawn of Everything: A New History of Humanity*.
33 Breggin, "The Biological Evolution of Guilt, Shame and Anxiety: A New Theory of Negative Legacy Emotions."
34 Sznycer et al., "Shame Closely Tracks the Threat of Devaluation by Others, Even across Cultures."
35 Stets and Carter, "A Theory of the Self for the Sociology of Morality."
36 Breggin, "The Biological Evolution of Guilt, Shame and Anxiety: A New Theory of Negative Legacy Emotions"; Sznycer et al., "Shame Closely Tracks the Threat of Devaluation by Others, Even across Cultures."
37 Barbalet, *Emotion, Social Theory, and Social Structure*.
38 Barbalet. P. 104.
39 Stets and Carter, "A Theory of the Self for the Sociology of Morality."
40 Kemper, *A Social Interactional Theory of Emotions*.
41 Clark, *Misery and Company: Sympathy in Everyday Life*.
42 Turner, *On Human Nature: The Biology and Sociology of What Made Us Human*.
43 Szanto, "Collective Emotions, Normativity, and Empathy: A Steinian Account."
44 Szanto and Moran, "Introduction: Empathy and Collective Intentionality—The Social Philosophy of Edith Stein."
45 Schwartz, "How Is History Possible? Georg Simmel on Empathy and Realism."
46 Stein, *On the Problem of Empathy: The Collected Works of Edith Stein*; Ferran, "Empathy, Emotional Sharing and Feelings in Stein's Early Work."
47 Decety, "Dissecting the Neural Mechanisms Mediating Empathy"; Decety and Cowell, "Empathy, Justice, and Moral Behavior"; Decety and Meyer, "From Emotion Resonance to Empathic Understanding: A Social Developmental Neuroscience Account."
48 Trevarthen and Aitken, "Infant Intersubjectivity."
49 Batson, "These Things Called Empathy: Eight Related but Distinct Phenomena."

50 Decety and Cowell, "Empathy, Justice, and Moral Behavior"; Rochat and Striano, "Social Cognitive Development in the First Year"; Panksepp, *Affective Neuroscience: The Foundations of Human and Animal Emotions.*
51 Humphrey, "The Uses of Consciousness."
52 Hommel et al., "The Theory of Event Coding: A Framework for Perception and Action."
53 Prinz, "Perception and Action Planning"; Decety and Jackson, "The Functional Architecture of Human Empathy."
54 Shepard, "Ecological Constraints on Internal Representation: Resonant Kinematics of Perceiving, Imagining, Thinking, and Dreaming."
55 Russell, *Agency and Its Role in Mental Development*; Decety, "Dissecting the Neural Mechanisms Mediating Empathy."
56 Batson et al., "As You Would Have Them Do unto You: Does Imagining Yourself in the Other's Place Stimulate Moral Action?"; Decety and Meyer, "From Emotion Resonance to Empathic Understanding: A Social Developmental Neuroscience Account"; Eisenberg, Spinrad, and Sadovsky, "Empathy-Related Responding in Children."
57 Reser and Swim, "Adapting to and Coping with the Threat and Impacts of Climate Change."
58 Legerstee, "The Role of Person and Object in Eliciting Early Imitation"; Hobson, "On Sharing Experiences"; Decety and Jackson, "The Functional Architecture of Human Empathy."
59 Denworth, "'Hyperscans' Show How Brains Sync as People Interact."
60 Hoffman, "Empathy and Prosocial Behavior."
61 Hoffman.
62 Tam, "Dispositional Empathy with Nature"; Geiger et al., "Observing Environmental Destruction Stimulates Neural Activation in Networks Associated with Empathic Responses"; Pfattheicher, Sassenrath, and Schindler, "Feelings for the Suffering of Others and the Environment: Compassion Fosters Proenvironmental Tendencies."
63 Milton, *Loving Nature: Towards an Ecology of Emotion*; Livingstone, "Taking Sustainability to Heart—towards Engaging with Sustainability Issues through Heart-Centred Thinking"; Perkins, "Measuring Love and Care for Nature."
64 Brown et al., "Empathy, Place and Identity Interactions for Sustainability"; Gutsell and Inzlicht, "A Neuroaffective Perspective on Why People Fail to Live a Sustainable Lifestyle."
65 Clark, *Misery and Company: Sympathy in Everyday Life*; Boehm, *Moral Origins: The Evolution of Virtue, Altruism, and Shame*; Guthridge et al., "A Critical Review of Interdisciplinary Perspectives on the Paradox of Prosocial Compared to Antisocial Manifestations of Empathy"; de Waal, "Putting the Altruism Back into Altruism: The Evolution of Empathy."
66 Hoffman, "Empathy and Prosocial Behavior"; Segal, *Social Empathy: The Art of Understanding Others.*
67 Decety and Cowell, "Empathy, Justice, and Moral Behavior."
68 Davis, Luce, and Kraus, "The Heritability of Characteristics Associated with Dispositional Empathy."
69 Lin et al., "Oxytocin Increases the Influence of Public Service Advertisements."
70 Decety and Jackson, "The Functional Architecture of Human Empathy."
71 McCaffree, "Towards an Integrative Sociological Theory of Empathy"; Thiermann, "Motivating Individuals for Social Transition_ The 2-Pathway Model and Experiential Strategies for pro-Environmental Behaviour"; Hoffman, "Empathy and Prosocial Behavior."
72 Hoffman, "Empathy and Prosocial Behavior."

73 Decety and Cowell, "Empathy, Justice, and Moral Behavior."
74 Bloom, "Empathy and Its Discontents."
75 Simas, Clifford, and Kirkland, "How Empathic Concern Fuels Political Polarization."
76 Ahmed, "Collective Feelings: Or, the Impressions Left by Others."
77 Hochschild, *So How's the Family? And Other Essays.*
78 Sapolsky, *Behave: The Biology of Humans at Our Best and Worst.*
79 Moon, "Powerful Emotions: Symbolic Power and the (Productive and Punitive) Force of Collective Feeling." P. 281.
80 Cikara and Bavel, "The Neuroscience of Intergroup Relations." P. 261.
81 Graeber and Wengrow, *The Dawn of Everything: A New History of Humanity.* P. 281.
82 Farmer and Maister, "Putting Ourselves in Another's Skin: Using the Plasticity of Self-Perception to Enhance Empathy and Decrease Prejudice"; Nussbaum, *For Love of Country: Debating the Limits of Patriotism*; Segal, *Social Empathy: The Art of Understanding Others.*
83 Christakis, *Blueprint: The Evolutionary Origins of a Good Society.*
84 Batson, "The Empathy-Altruism Hypothesis: Issues and Implications."
85 White, *Fearful Warriors: A Psychological Profile of U.S.-Soviet Relations.*
86 Breggin, "The Biological Evolution of Guilt, Shame and Anxiety: A New Theory of Negative Legacy Emotions."
87 McCaffree, "Towards an Integrative Sociological Theory of Empathy"; Blau, *Inequality and Heterogeneity: A Primitive Theory of Social Structure*; Blau, *Structural Contexts of Opportunities*; Blau and Schwartz, *Crosscutting Social Circles.*
88 Tokita, Guess, and Tarnita, "Polarized Information Ecosystems Can Reorganize Social Networks via Information Cascades."
89 Krznaric, *Empathy.*
90 Stewart, Plotkin, and McCarty, "Inequality, Identity, and Partisanship: How Redistribution Can Stem the Tide of Mass Polarization."
91 Bednar, "Polarization, Diversity, and Democratic Robustness."
92 Ruiz-Junco, "Advancing the Sociology of Empathy: A Proposal."
93 Gilligan, *In a Different Voice: Psychological Theory and Women's Development.*
94 Sapolsky, *Behave: The Biology of Humans at Our Best and Worst*; Piff et al., "Higher Social Class Predicts Increased Unethical Behavior."
95 Sapolsky.
96 Morris, "Empathy and the Liberal-Conservative Political Divide in the U.S."
97 Stewart, Plotkin, and McCarty, "Inequality, Identity, and Partisanship: How Redistribution Can Stem the Tide of Mass Polarization."
98 Jamieson, "Sociologies of Personal Relationships and the Challenge of Climate Change."
99 Rizzolatti and Sinigaglia, "Mirror Neurons and Motor Intentionality."
100 Coviello et al., "Detecting Emotional Contagion in Massive Social Networks"; Nakahashi and Ohtsuki, "When Is Emotional Contagion Adaptive?"
101 He et al., "Exploring Entrainment Patterns of Human Emotion in Social Media." P. 2.
102 Summers-Effler, Van Ness, and Hausmann, "Peeking in the Black Box: Studying, Theorizing, and Representing the Micro-Foundations of Day-to-Day Interactions"; Iacoboni, *Mirroring People: The New Science of How We Connect with Others.*
103 Rizzolatti and Sinigaglia, "Mirror Neurons and Motor Intentionality." P. 192.
104 Denworth, "'Hyperscans' Show How Brains Sync as People Interact."

105 Fowler and Christakis, "Dynamic Spread of Happiness in a Large Social Network: Longitudinal Analysis over 20 Years in the Framingham Heart Study."
106 Coviello et al., "Detecting Emotional Contagion in Massive Social Networks"; He et al., "Exploring Entrainment Patterns of Human Emotion in Social Media."
107 Iacoboni, *Mirroring People: The New Science of How We Connect with Others*.
108 Christophe and Rimé, "Exposure to the Social Sharing of Emotion: Emotional Impact, Listener Responses and the Secondary Social Sharing"; Heath, Bell, and Sternberg, "Emotional Selection in Memes: The Case of Urban Legends."
109 Downs, "Prescriptive Scientific Narratives for Communicating Usable Science"; Cormick, "Who Doesn't Love a Good Story?—What Neuroscience Tells about How We Respond to Narratives."
110 Durkheim, *The Elementary Forms of the Religious Life*.
111 He et al., "Exploring Entrainment Patterns of Human Emotion in Social Media."
112 McCoy and Major, "Group Identification Moderates Emotional Responses to Perceived Prejudice."
113 Rimé, "Emotions at the Service of Cultural Construction."
114 Goffman, *The Presentation of Self in Everyday Life*; Hochschild, *The Managed Heart: Commercialization of Human Feeling*; Langford et al., "Social Modulation of Pain as Evidence for Empathy in Mice."
115 Verweij, "Emotion, Rationality, and Decision-Making: How to Link Affective and Social Neuroscience with Social Theory"; Turner and Stets, "Sociological Theories of Human Emotions."
116 Sapolsky, *Behave: The Biology of Humans at Our Best and Worst*.
117 Turner and Stets, "Sociological Theories of Human Emotions."
118 Sapolsky, *Behave: The Biology of Humans at Our Best and Worst*; McCoy and Major, "Group Identification Moderates Emotional Responses to Perceived Prejudice."
119 Alshamsi et al., "Beyond Contagion: Reality Mining Reveals Complex Patterns of Social Influence."
120 Amel et al., "Beyond the Roots of Human Inaction: Fostering Collective Effort toward Ecosystem Conservation."
121 Jasper, "The Emotions of Protest: Affective and Reactive Emotions in and around Social Movements." P. 398.
122 Amel et al., "Beyond the Roots of Human Inaction: Fostering Collective Effort toward Ecosystem Conservation."

5 Scaling Up Emotions, from the Individual, to Social Structures and Back Again

Why talk about institutions and social structures in a book focused on emotions? We often view emotionality as a private affair, and the lion's share of research in the affective sciences tends to abide by this view: the previous chapter largely focused on emotionality at the individual level. However, most of those same researchers readily acknowledge that, while rarely the focus of their research, our inclination as social creatures to share emotions means that there are layers above and beyond the individual in which we can observe emotionality and its role in social change.[1] These relational components of emotionality encompass emotional responses to perceptions of equity, and pursuit of social belonging, among other things.

The imprint of our emotionality does not stop at the level of social interaction, however. Social structures generate conditions that provide feedstock for our emotional responses, while at the same time impose rewards and sanctions on that same emotionality. Our emotional experiences are deeply entangled with our socio-cultural and political context, even though we may be partially or wholly unaware of those entanglements, so omnipresent are they in our lives. Social structures and the institutions that embody them constitute part of the environment that makes up our emotional climate[2]—the other part being our direct, embodied relationships with the nonhuman universe—to which we each respond affectively, and which, through experience and socialization, shape our emotional responses to events. In fact, if it were not for our emotionality, we would be unlikely to create social structures at all. Of course, the problem is, once created, social structures can become rather overbearing entities in our lives that are resistant to change, even when they no longer serve us well, and never did serve some of us. While we humans may be their creators, social structures turn around and re-create us. They affect our emotionality by prescribing social identities and affiliated behavioural norms and sanctions. Our emotionality in turn guides our negotiations with those structures, as we each reflect upon, and navigate them along emotion-cognition pathways that guide some of us towards deference, and others towards collective commitments to change.

As discussed in Chapter 3, social structures are a rather elusive concept, but they are manifest in something far more concrete—institutions—through which the patterns and beliefs prescribed by those social structures operate. Many of the effects of social structures—material, emotional, and cognitive—are specific to institutional settings. Others, however, can be traced to macro-social structures that, while further obscured from our vision, thread their way through *all* institutions within a given social system. Within each of the institutions operating in a given social system, for example, we are likely to observe similar norms and practices pertaining to gender and race, which reflect how gendered and racialized relations are prescribed within those social systems.

One particularly glaring set of macro-social structures, the dinosaur in every boardroom and classroom, on every street corner and farmer's field across the globe today, is capitalism, and its bedfellows, colonialism, and patriarchy. These are three social structures that I consider to be particularly germane to social relations with our climate, as they thread their way through the economic, scientific, educational, and government institutions directly engaged in our interactions with the natural world. While there are regional variants to be sure—patriarchy looks different in the Middle East than it does in North America, for example, and on Wall Street compared to the Corn Belt—this triadic structure has spawned a set of hegemonic ideologies: consumerism, individualism, sexism, racism, and human exceptionalism, all of which have served to enable the exploitation of climate and ecosystems, while at the same time compromising the collective capacity to confront that exploitation. Much has been written about the ills of late capitalism, particularly for ecological and human wellbeing. Less discussed, however, is the means by which capitalism operates upon our emotionality. What does living with capitalism look *and feel* like, for that white male CEO, for middle-class single moms, for Black teens living and schooling in under-served communities, for Indigenous Elders upon whose lands we occupy, dig, build, and dump? Can any of us, embodying unique positionalities yet living under the same capitalist roof, imagine life without it?

How Social Institutions Impose upon Our Emotionality, Doing So in Deeply Intersectional Ways

Before delving into those social structures, let's unpack what our institutional context means for emotionality. Institutional norms and practices are not inherited, pre-programmed traits; because of their cultural origins, they must be learnt, and each of us has the capacity to deviate from them. Thus, if they are to persist, they require active support, through emotion-triggered rewards and sanctions. The shared norms and meanings imposed by institutions channel individual and collective sentiments and actions, by deeming certain behavioural choices appropriate, and others not.[3] This includes emotional behaviour; the more one behaves emotionally in a manner expected, the more one is said to express 'emotional competence,'[4] which can require at times a strong dose of personally taxing emotion management—eating for lunch that panoply of emotions that are deemed inappropriate to given situations, and spewing emotional performances that do not match our inner feelings. However, while emotions can be managed, they cannot simply be managed *away*. Rather, we end up with what Arlie Hochschild[5] calls *emotive dissonance*: analogous to cognitive dissonance, emotive dissonance refers to a disconnect between emotions felt and emotions expressed, which can generate a profound alienation.

Emotions are the motivational force for what Moon[6] refers to as symbolic exclusion: efforts to assert performatively the boundaries of the institutional collectivity, by filtering information to avoid new understandings and alliances that threaten institutional cohesion, and ensure compliance. Some institutions may prescribe formal rules, such as professional codes of conduct, but for the most part an institution's members themselves partake in this policing. We each have engaged in sanctioning others for deviance in subtle and not-so-subtle ways, from teasing someone about their wardrobe, to expressions of disgust towards others for their actions, or for who they are (homeless, Black, Muslim, transgender). This includes people in positions of power and influence in particular—managers, journalists, teachers, coaches—but such active policing does not stop here. Those people who embrace strong levels of emotional commitment to a given institutional ethos become particularly active boundary police, shaming deviance as it arises[7]; in fact, active participation in such policing might seem like a good way to improve one's own status. Through socialization, we become quite effective self-policers, either in response to our desire for approval, or—often an eventuality—because those norms and meanings become internalized. For

Douglas Creed and colleagues,[8] systemic shame and episodic shaming are forms of disciplinary power that operate both through surveillance and self-regulation. Avoidance of shame is sought through conformity; success in doing so invokes shame's opposite—pride. However, for many who lack the capacity to conform to expectations, that shame can become systemic, with enormous consequences for personal agency and mental health. The disciplinary power of institutional norms can become so internalized that they escape our consciousness, to such an extent that it can become difficult for those in marginalized positions to recognize that a given institutional ethos does not serve their interests, and in fact serves to maintain their marginality. Many of us may in fact become active adherents of an institutional ethos that compromises our own wellbeing.[9] Socialization thus has a self-perpetuating tendency, oppressing our capacity for reflexivity and agency, which is one of the most important reasons why institutions are so resistant to change.

Elites are also capable of manipulating emotionality, intentionally or circumstantially, to favour certain behavioural outcomes, or at least, to capitalize upon them, such as elevating alarm over perceived terrorist threats in order to elicit support for police states.[10] Many structural conditions also produce different emotional effects according to one's intersectional positionality. The outcomes for individuals living under patriarchy, for example, have distinct effects for men and women; colonialism for settlers and Indigenous peoples. As such, disciplinary power transpires in the differential distribution of emotion states in a given population.[11] The more privileged an individual's station, the more likely they are to feel pride, entitlement, and efficacy. The less privileged one is, the more an individual is likely to experience systemic shame, and also uncertainty and fear—the fear that comes with powerlessness. As Turner[12] articulates, our inclination to abide by dominance, territorial control, hierarchies, and the inequities that ensue are by no means 'natural' or inevitable features of human societies. To the contrary, these 'cages of power,'[13] which result from ever-more oppressive norms and sanctions associated with ever larger and more complex institutional settings in modern societies, conflict with our inherited nature in many ways. We all have an inclination to rebel against such cages, and gravitate towards those social networks, institutions, and societies that allow for the expression of our need states. Or, in the absence of such alternative social spaces, we instead are inclined to escapism, poor mental health, and self-destructive behaviours.

Some Things to Be Said about Capitalism

Let's first name the elephant. Despite its varying forms, there are five truths that I consider to be elemental to modern capitalism, in all of its varying forms. The five truths of capitalism I describe below did not suddenly appear the moment capitalism emerged from its feudal and mercantilist parents. These five features arguably did not achieve full fruition until the introduction of communication and transportation technologies that enabled rapid globalization in the post-World War II era, and the enthusiastic adoption by Western political leaders of neoliberalism in the 1970s. Indeed, the neoliberal, global capitalism of the 21st century bears little resemblance to the economic system envisioned by its early ideological craftsmen, like Adam Smith, who premised his theories on face-to-face, interpersonal commerce among community members. As noted by Rollert,[14] Adam Smith was an astute psychologist, who recognized that acquisitions were not ends in themselves; rather our engagement in the marketplace reflected our desire for social belonging—the same driving force that allows the creation of institutions. The anonymized interactions typical of the globalized capitalist marketplace today could not look more different. Now, let's get to those five truths.

First, capitalism's modus operandi is the production of profit, and because profit represents the difference between the capitalist's investment into production and the prices received for the product, the generation of profit can only be realized through exploitation of the means of production, of which there are really just two types: people and planet. Planetary exploitation continues to advance, as capital seeks its inputs and sacrifice zones ever further afield, applying technologies and practices that allow for continued intensification at the same time as the capacity of the land to provide material inputs and absorb waste diminishes. 'Capitalism's governing conceit is that it may do with Nature as it pleases,' says Jason Moore.[15] And while states are to varying degrees called upon by certain constituencies to intervene to manage the disruptions this entails, without the ability to exploit, there would be no profit, hence no capitalism, and since states in capitalist societies are as addicted to profit as capitalists themselves, most nation-states in capitalist countries tend to toe the line.

The second involves the delineation of societies into hierarchical orders of owners, workers, and dependants, which fortifies rigid systems of inequality on the basis of access to wealth and manifests in political power. This generates endemic tensions and conflicts in those social systems. Such hierarchies are by no means one-dimensional, however. To the contrary, they are intersectional. On the one hand, white supremacy has served as an allocation metric whereby non-white peoples are deemed of lower status, justifying their economic marginalization or outright enslavement.[16] Children, elderly, and disabled are, furthermore, devalued by virtue of their unemployability. Women and Indigenous peoples, to be discussed in greater detail later, are likewise marginalized and oppressed. These constructions of categories of difference between peoples—between white and non-white, men and women, settler and Indigenous—provides an ideological curtain behind which brute enforcement of inequities justify the accumulation of wealth among the few, who also happen to be white and male by and large. The resulting intersectional hierarchy serves to divide at the same time as it lumps—there is, for *almost* everyone, someone else above and below, generating a 'matrix of domination,'[17] and thus relations among each other are rendered competitive, rather than cooperative. Marginalized peoples are in varying ways indoctrinated into the service of producing wealth for others. The homes and neighbourhoods of those very servants then become the dumping grounds for waste,[18] and their bodies become the proverbial and—all too frequently—actual punching bags to allow for the release of frustrations felt by all whose dreams and expectations manufactured by capitalism remain unrealized, indeed unrealizable.

Third, profit generation requires continuous consumption, which means continuous demand for commodities. And because the satiation of basic needs, reliance on subsistence, commons systems and sharing economies, and the long-term use of durable goods all amount to a reduction of demand, these things too are the enemies of capital. Therefore, an additional modus operandi is the creation of demand, by continuously creating new markets through the dismantling of local and subsistence-based economies, and by ensuring that those with the means to consume are never sated, by reducing the subjective and real exchange value of commodities from the point of purchase, either with planned obsolescence or by creating new 'needs' through marketing. Consumption becomes the ticket to access social networks, with different commodities allowing access to different networks. We are identified by—we *are*—what we consume. To this end, our emotionality, our need to belong, becomes our own worst enemy. Capitalism serves to redirect our pursuit of belonging in the form of consumerism, and yet that pursuit is one that is never accomplished. Our home is never grand enough, our clothing always out-dated, our wealth objectives ever expanding. When one's identity and status are established through displays of consumption, ultimately one's latitude in their pursuit of self-identification is determined entirely by one's access to wealth.[19]

Fourth, and relatedly, capitalism supports individualization. While it would be rash to suggest that capitalism 'caused' individualization, the relationship has most certainly been facilitative. Capitalism is a force of individualization of interests, a prescription for selfishness, and for the privatization of commons, as with the violent—to peoples and the land—privatization and division into individual allotments of Indigenous lands to encourage European settlement across the West. Or the equally violent incursions of mining in Yanomami territory in the Brazilian Amazon happening today. Capitalism is a community killer; it defines each of us as 'our own boss,' while ensuring only a small minority are actually bosses, and instilling competitiveness and distrust between us. Strong communities are sites of resistance after all, as is civic engagement of all forms. Kill civil society, and the sites of resistance to capitalist expansion are killed with it. With the help of economists and their rational-utilitarian model of human behaviour, individualization is reinforced by the academy, and by Western legal and justice systems.

Fifth and finally, the trick, the key to the perpetuation of capitalism, is to never let the cat out of the bag; to insist, over and over, that what is good for capital is good for the planet and good for you and me. This master narrative requires a vast effort in information management. And to an even greater degree, emotion management. Never reveal the exploitation and the manipulation that ensure the protection of the wealth of the few. As Jurgen Habermas[20] notes, the greatest challenge for contemporary societies is to allow for the distribution of resources in a manner that is highly inequitable yet viewed as legitimate. The grand narrative we have received, from religious leaders, from elected officials, from the learned societies, corporate spokespersons, the media, and our school teachers, is that capitalism is the great enabler of freedom and enterprise, and your impoverishment is your own failing. The many costs of overconsumption are similarly the consumer's doing: *you* chose to smoke, drink, and consume processed foods. Companies are in the business of meeting our demands, and thus capitalists are merely in the service of the consumer, or so we are told. In fact, what we need in order to address crises of production and consumption as they arise is more capitalism.

Our emotionality offers capitalism one of its most useful sources of stability (and also its fragility—more on that later). Consumer demand depends upon the creation of desires. However, if the desire that is cultivated is not actually for things themselves, but rather the positive emotions associated with the social status and group belonging that those things offer, then the things themselves—the commodities—that allow for such status and belonging can change continuously. Our emotionality also becomes the means to maintaining quiescence and complicity, in the workplace, in public spaces, in the private sphere. As much as our consumer desires are facilitated by emotions, the policing of emotionality in a manner that moulds us into rational, self-interested consumers, compliant and productive workers, and passive supporters is achieved through emotional repression; the cultivation of what Theodor Adorno called 'bourgeois coldness.'[21] According to Jo Ann Pavletich,[22] emotional control was prescribed as early as the Victorian era. The dominant European culture at the time developed a strongly negative attitude towards intense emotions, seeing such expressions as antisocial. One side effect of this was a distinct masculinization of culture. Sentimental aspects of culture, thus feminized, lost their legitimacy. For Frank Weyher,[23] 'capitalism prescribes the "rationalizing" and "intellectualizing" of all relations and a deceptive "limiting" of the range of "emotional" life, sanctioning those emotions that conflict most obviously with "reason," particularly the "instrumental" forms of reason most consonant with capitalism.' This emotional control also enables our apathy towards the plights of those below us. I don't want to see or hear about your distress! The less I have to see, hear and feel your distress, the more I am able to continue to pursue my self-interest. We can spare no time to grieve, distress and sadness must be medicated away, and under no circumstances can we find joy in things that are not commodifiable. This flattening of our emotionality, ironically but beneficially for capital, translates into even more consumption, as we seek emotional gratification and a sense of autonomy through a commodity-driven culture.[24]

Capitalism's Emotional Toll

While we all within capitalism experience Hochschild's emotive dissonance to some degree, none so much as those occupying the lowest rungs. Many service industries, for example—dominated by women—require emotion labour: the necessary painted-on smile, the patience, the effort to make someone else feel supported, the attentiveness to another's needs, regardless of what that worker is really feeling in that moment. In these sectors, for Hochschild, the labourer's emotions are just one more element of social life that becomes commercialized.

This move towards emotional self-responsibility and self-control and its resulting emotive dissonance can create conditions that precisely lead to *increases* in the intensity of certain negative emotions, due to the loss of relationships, of emotional sincerity, and of ontological security—referring to the comfort that comes when one knows one's place and what to expect—and when one's beliefs are validated. Personal responsibility for achieving the unachievable, for example, is linked with fear and shame. Long-term precariousness generates a loss of motivation and hope for the future. These effects have been exacerbated under late capitalism's withdrawal of welfare states and labour unions, the increasing precarity of employment, and spectacular increases in economic polarization.[25] The manipulation of emotions under capitalism has enormously impactful personal effects—poor mental health being one—and social effects, including polarization, and the enthusiastic uptake of disinformation, or 'alternative' facts. While there are many mechanisms at work here, the escalation of fear and inhibition of empathy are among the most consequential.

Up with Fear

Zygmunt Bauman[26] called fear an enduring symptom of late capitalist societies. We have a limited ability to empirically substantiate the claim by Bauman and other modernization theorists that modernity is marked by a generalized increase in fear in comparison to previous eras, and it is most certainly the case that the nature and degree of personal experiences of fear vary starkly. There are nonetheless numerous compelling arguments and observations that living in late, modern, capitalist societies today is marked by an undercurrent of ontological insecurity which, lacking the resources to re-establish an

90 *Scaling Up Emotions*

ontologically secure state, tends to generate fear. Fear and anxiety increase when one's own power decreases, and late capitalism has usurped the power and autonomy of the vast majority. Control (e.g. of workers by bosses, of citizens by regulators) serves to reproduce structural power, and is achieved through the exploitation of fear and anxiety.[27] For Jonathan Turner,[28]

> authority structures ... are shame-generating machines, and when coupled with the unequal distribution of valued resources, they increase the likelihood that large numbers of individuals will not meet expectations for resources—income from jobs, prestige from educational credentials that give access to money and power, love from family, or power from unions or political parties.

What does fear do to a person? It narrows our gaze, increases the production of stress hormones, and escalates feelings of distrust and self-protection. These emotional pathways can then become self-reinforcing, with fear motivating actions the results of which include enhancing conditions that produce fear, by, for example, allowing distrust to weaken current social bonds and resist new ones, or failing to heed warnings that come from distrusted others, including, say, scientists and journalists. Martin[29] identifies several major developments that tend to generate fear in late capitalism. The first involves globalization, which results in the rupture of stability provided by local social and cultural contexts.[30] The resulting ruptures, including shifts in hierarchical positions and occupational roles, have disintegrated traditionally stable identities and lifestyles, imposing greater demands on one's personal reflexivity, as many prior 'givens'—Where will I live? What should I do?—are no longer to be taken for granted.[31] In addition, the global risks introduced by the pursuit of profit via industrial and technological means, including global warming of course, but also chemical contamination, nuclear accidents—what Ulrich Beck[32] refers to as the latent side effects of capitalism—loom large in the fears of many living in the late 20th and 21st century. While the fear-generating effects of each of these developments is not inevitable, individualization strips us of one of the most important resources we have for confronting challenges in order to quell our fears: our social relationships. Just how acute are the personal consequences of these fear-inducing developments depends entirely on one's intersectional positionality. Not all members of contemporary society are equally subjected to precarity of employment to the same degree. Many but not all of us are saddled with additional intersectional burdens: racism, sexism, Islamophobia, queer-phobia, and so on.

The deregulation and individualization of lifestyles, in turn, supports a fear-related politics of inequality. As Sara Ahmed[33] reminds us, 'fear is ... dependent on particular narratives of what and who is fearsome.' The fears of Americans after 9-11, for example, were embedded in Islamophobia.[34] Fears held in common can produce in-group bonds that create the basis for social collectives—think working-class white men—that are reinforced by attacking feared Others, as we are seeing today with the rise of right-wing populism[35] and the practice of fear-invoked techniques of governance such as surveillance[36] directed at social minorities: immigrants, refugees, sexual minorities, or the mentally ill. As described by Sara Ahmed,[37] late capitalism generates fears of Others which when politicized favour the reproduction of social inequalities that in turn further enhance those fears.

Down with Empathy

In comparison to the extensive attention to fear, relatively few researchers have discussed capitalism's impacts on empathy. However, that individualization discussed by Bauman and others, and the bourgeois coldness Adorno spoke of, are in effect compromises to our capacity for empathy. Bourgeois coldness may well support capitalism, but it also feeds a personal sense of emptiness, a lack of passion and fulfilment, and compromised ability to cultivate and maintain relationships, leading to what Adorno called an 'atrophy of lived experience'[38] that becomes a vicious cycle.

Our basic predispositions for empathy are also compromised by the compartmentalization of our social universe by economic status. Frequency of social interaction remains the most likely inducement for intersubjective understanding, but capitalism locates people in social compartments defined by wealth, income, and occupation, which determine the likelihood for social interactions between individuals to occur. As discussed in Chapter 4, fluidity of group membership allows for cooperation and altruism to emerge across group boundaries. However, if social structures reinforce those boundaries, then such fluidity is hampered. Living under such conditions of stark inequity imposes barriers to empathy in other ways. As noted in Chapter 4, inequity compromises the efficacy and mental health of marginalized groups though repeated confrontations with environmental contamination, discrimination, and colonialism.

Emergent Effects of Capitalism's Emotional Toll

As independent as they may appear, the tandem emotional impacts of capitalism on fear and empathy are deeply entangled, and the emergent outcome is a social world of conflict and distrust, while stripped of our inherited capacities to combat that conflict and distrust—to cooperate. The ultimate effect is a stifling of the very emotionalities and collective capacities that constitute the necessary precursors to resistance. Highly inequitable societies that allocate stark differences in privilege, on the basis of material wealth but also highly racialized, gendered, and colonial societies, describe an important set of compartmental boundaries to our expressions of empathy. Inequity is also linked to increases in affective (as opposed to ideological) polarization, referring to experiencing negative affect when thinking about Others.[39] Politics in the

21st century have entered into a period of intense polarization,[40] driven to a great extent by fear and distrust.

The introduction of social media into such a social landscape has been nothing short of explosive, including the profligate availability of disinformation, and the fortification of social divisions through echo chambers.[41] Petter Törnberg[42] explains: when social interaction takes place in local social networks, there tends to be limited sorting on the group level. Some preferences are politicized in one region but not in others, serving as a check on political polarization and creating dialogic space for cross-cutting initiatives, and relatively high levels of social cohesion. Social media, however, allows global alignment of preferences, which then become hardened through lack of interaction with others, particularly involving social identities such as nationality, ethnicity, language, religion, and gender.[43] These are social identities in which solidarity and belonging with other group members can still be experienced in the framework of shared concerns, emotions, and meanings, because they do not involve competition in the capitalist order.

These new alignments create space for extremist positions,[44] support for which is cultivated by—you guessed it—appeals to emotions, like nostalgia for mythical pasts, fear of Others,[45] and anger and resentment. Recent research suggests that persistent anger can weaken commitments to democratic values—particularly our respect for those with whom we disagree,[46] in some cases going so far as to dehumanize opponents, eradicating any potential for deliberation.[47] The re-grouping allowed by social media in effect offers a means to translate personal vulnerability and shame into anger and blame.[48]

Patriarchal Underpinnings of Capitalism

There is of course a need for empathy, for care and nurturing, to the functioning of societies, but capitalism has solved this puzzle by relegating this responsibility to women. While men are tasked with production, women—including both their labour and their bodies—are brought into the service of capital by delegating to them the responsibility for social reproduction: the feeding, clothing, protecting, healing, and caring labour necessary to the raising of families and communities, generating the responsible and healthy adults and cooperative workers that allow for the perpetuation of economic growth.[49] When women commit to doing all the emotionally intensive care work necessary to maintain stable and healthy families and communities, men don't have to.

Barring slavery, some other means of compelling such labours is required. Few of us would be inclined to graciously accept oppression, after all. This work is done quite effectively by patriarchy. Patriarchy has surfaced in various forms in many places across the globe throughout history, and thus cannot be said to have emerged strictly to serve the needs of capitalism, and a capitalist system in the absence of patriarchy is at least hypothetically imaginable. Nonetheless, capitalism and patriarchy clearly enjoy a synergistic relationship. For Ferguson and colleagues[50]:

> control and degradation are secured concretely in, and through, the negotiation of race, gender, sexuality, and other layered and interwoven social relations. These are the relations which ensure that labour arrives at capital's doorstep ready to be further dehumanized and exploited.

For bell hooks, in her essay *Understanding Patriarchy*, patriarchy is a system in which 'males are inherently dominating, superior to everything and everyone deemed weak, especially females, and endowed with the right to dominate and rule over the weak and to maintain that dominance through various forms of psychological terrorism and violence.' Patriarchy is one of many elusive and multi-definitional sociological concepts (is there any other kind?), but most treatments refer to patriarchy as a set of social structures, encompassing modes of production, relations in the workplace, a complicit state, sexual relations, condoned male violence, and culture.[51] The over-arching logic of patriarchy is defined by a set of constructed dualisms between men/masculine and women/feminine, which assign superiority to all attributes associated with the former, and inferiority to the latter. Men are thus afforded autonomy and selfhood; women are passive and objectified. Men become the agents of history, women part of the furnishings. In a patriarchal society, men produce, women reproduce. Men govern, women abide. Men lead, women follow. Men think, women feel; men are strong, women are weak, not just physically but in mind and spirit as well. In the presence of racism and colonialism, moreover, the privilege meted by patriarchy is restricted to able-bodied, straight, white, Western men, while disabled, non-white, non-western, non-hetero, and non-cisgendered men are, to varying degrees, placed on the female/feminine side of the ledger, producing complex hierarchical orders among the subjugated.

Acquiescence Ensured through Gender Norms

Social systems characterized by subjugated relationships are inherently unstable, and thus cultural prescriptions are necessary to maintain the legitimacy of what is an illegitimate system, to avoid revolt. In patriarchal capitalist systems, 'women's work' involves maintaining the conditions of reproduction, but since this work does not generate surplus value, this work is not remunerated. Even since these services have become commodified under neoliberalism, paid care work is typically precarious and low-paying, perpetuating economic dependence. The commodification of such roles under neoliberalism has generated 'global care chains,' as nonwhite women and men from the Global South are imported into the service of Global North men, families, and communities, sometimes through legal means, in others by trafficking, including everything from nannies and housework, to mail-order brides and sex tourism.

To ensure the stability of the system, to ensure that these essential services are provided, women must be compelled by other means than remuneration to perform these services for capital. This is achieved through the assignment of gendered norms; what Gilligan and Snyder[52] call 'codes of manhood and womanhood' that prescribe appropriate behaviour, behaviour that upholds masculine honour and women's servitude. While the specific prescriptions vary by cultural context, in general, women who are too assertive, too sexual, inattentive mothers, or otherwise noncompliant are sanctioned and made to feel shame for stepping outside the bounds.[53] Men, in contrast, are rewarded for the exact same behaviours but sanctioned for others, such as developing close bonds with other men, exhibiting physical weakness, or expressing care and vulnerability. Sexual orientations inevitably become implicated in this scheme, as non-hetero-normative orientations are by definition violations of gender norms of masculinity and femininity.

Emotional Costs of Patriarchy

The effort to abide by gender norms involves an emotional toll on everyone, in different ways, but in particular for all those who fall onto the subjugated side of the ledger.[54] We have evolved to be relational creatures: equitable, reciprocal, empathic, and connected, while also retaining our self-hood, our agency. These attributes are key to personal fulfilment, and by extension to our ability to confront collective problems. It simply feels good to exercise the capacities—for emotion, for caring, for autonomy, that we inherited. Patriarchy seeks to break that down entirely in order to allow for power, privilege, and stature among a handful.

Among men, who are assigned the expectation of domination and control over others, control necessarily requires a certain dismissiveness. Men's inexpressiveness is a prerequisite for assuming positions of power and privilege, in order to be able to avoid feelings of empathy towards those who are affected by their actions.[55] Discouragement of personal attachments is enormously costly, producing what Jonathan Rutherford[56] refers to as men's emotional illiteracy, manifesting in feelings of loneliness and isolation. Even those who have achieved the highest levels of success that patriarchy has to offer can end up leading 'existentially empty lives, producing wealthy "hungry ghosts" whose appetites are never sated.'[57]

However emotionally contained some men may appear to be, notions that men are 'unemotional' are false. Men under patriarchy are allowed a highly emotional expressiveness in certain highly restricted domains. Pride, for example, is the reward for pursuing hegemonic masculinities, and men are encouraged to express their pride, in everything from their occupations and nationalities, to their wives and their trucks. And while women are punished for expressions of anger, men are seen as entitled to theirs,[58] and even violent expressions of male anger are condoned as justifiable when their dominance, their privilege, are threatened.[59]

The emotional toll of patriarchy on women and subjugated Others takes two forms. The first is external, involving the assignment of emotion labour onto women. Across multiple occupational contexts, women are herded into emotion-giving/care-giving roles, while at the same time women are still expected to go home after the workday and bear the primary burden of childcare and household management. Men thus freed of emotion labour can immerse themselves in cognitive and abstract pursuits. The second is internal: the personal emotional toll of subjugation.[60] While women are encouraged to express caring emotions, all forms of resistance or rebellion to their subjugation, including anger and rage, are sanctioned. Instead, women are encouraged to stay silent, stripped of their agency and their rights to self-care, even in the face of rape and violence.[61]

Men are told not to express care, not to express vulnerability. Women are told that their only purpose is to care—for anyone other than themselves. Men lose their relationality; women lose their selfhood. Failure to live up to these expectations, however, induces estrangement and shame. Therefore, we carry on, by stuffing those very emotional responses that also happen to have the potential to challenge capitalism. In the words of Ruba Ali Al-Hassani,[62] '[a]s long as men equate violent domination and abuse of women with privilege, they do not realize the patriarchy's damage to themselves and others, and do not rebel against it.' Gilligan and Snyder[63] agree:

> Patriarchy persists because it sanctions against the very beliefs and practices that would bring about its own demise ... in order to defend against a loss that has come to seem irreparable, we denigrate and detach from those very relational capacities necessary for repairing the ruptures that patriarchy and all forms of hierarchy create.

Patriarchy and the Planet: As Women Are Treated, So Too Is Nature

The justification for the subjugation and control of women under patriarchy largely rests upon the determination that women are 'closer to nature.' Little wonder then that under patriarchy the planet—including our many planetary systems and all of its nonhuman inhabitants—is likewise viewed as an object to be dominated and controlled. This is the key argument posed by ecofeminist scholars: patriarchy defines relations between men and the planet according to the same rubric that it defines relations between men and women. Mary Mellor[64] puts it this way: 'Industrialism, capitalism, colonialism, racism, and patriarchy are the different manifestations of a many-headed Hydra that has its fingers around the throat of women, poor people, and the planet.' Gender oppression and environmental destruction are thus intertwined according to Johanna Oksala[65]:

the feminization of nature and the naturalization of women do not function merely as ideological justifications for an abstract and general logic of domination, but concretely structure the capitalist society through gendered social and economic practices and divisions of labor.

The generation of surplus value requires exploitation, and patriarchy simultaneously provides the ideological enablement that allows for both the expropriation of women's reproductive and care labour and the exploitation of land, air, water, and nonhuman species.

Colonial Underpinnings of Capitalism

The history of capitalism did not begin with the industrialization of mass production; capitalism was borne much earlier, with the first travellers from the Empire who commodified lands, peoples, and their fruits in places that were not their home. Quite simply, without colonialism, there could be no capitalism. Colonialism involves the appropriation of lands and the resources they contain. Since these lands are always already occupied, have been lived with and loved for centuries before the first ships sailed away from the so-called Old World, the peoples occupying these lands must thus be moved aside, by whatever means of extermination or containment necessary, encompassing a myriad of creative physical, economic, legal, and psychological processes.

Referring to colonialism as a singular phenomenon is problematic, given the many forms it takes. Regions subject to colonialism today include, for example, colonies in which colonial rulers have officially handed over the rights to rule to a 'post-colonial' government, such as in the Democratic Republic of Congo, or remain territories under the control of the ruling state, such as Puerto Rico. Still others, like Canada, are more complicated Settler states, in which Indigenous and Settler peoples co-occupy the same lands, transpiring in daily confrontations and co-mingling of the past with the present. Each of these regions, in turn, feature distinct socio-cultural, legal, and economic structures facilitating and endorsing colonialism. But the objective remains the same: control over the wealth-enhancing potential of Indigenous lands, which requires the subjugation of their peoples.

Historically, colonialism has manifested as military projects, and in many parts of the world this remains the case, although overt colonialism has become officially uncouth. The subjugation of Indigenous lands and peoples continues nonetheless, under the more subtle practices of assimilation, erasure, paternalism, and other means, in an ever-adapting, contemporary "shape-shifting" neo-colonialism.[66] One colonial force with enormous consequences entails the ecological destruction of Indigenous lands by processes both local (mining) and global (global warming), a process Jules Bacon[67] refers to as colonial ecological violence. This may include environmental contamination of land, water, and air, and resource depletion, including those resources essential for subsistence such as fish, wildlife, and other harvestables. It also includes, sometimes intentionally, the desecration of sacred places. Ironically, efforts by state and non-state organizations to protect ecosystems can also manifest as colonialism, when Indigenous peoples are forced to refrain from traditional subsistence activities, or are forcibly removed from their lands entirely, in the name of conservation.

Although many of European descent insist colonialism is a subject for the history books, Bacon[68] reminds us that 'its traces can be found across all levels of analysis from the international to the interpersonal,' not only throughout history, but today. Colonialism is not just history, because it is bound with capitalism's treadmill: as soils and resources are rapidly exhausted in one region, ever more incursions into Indigenous territories are necessary to continued capitalist production. Schild[69] catalogues a handful of recent colonial exploits, including in Peru where, by 2013, 45 percent of Indigenous territories had been provided as concessions to mining operations. A sum total of 75 percent of the Amazon region—once heavily populated by Indigenous peoples that were decimated over the previous 400 years but whose descendants continue to rely on this rich ecosystem today—is controlled by oil consortia. In Guatemala alone, 500,000 Indigenous people had been expelled from their homes by 2012, to make way for mineral extraction. In Canada, a nation-state that has formally signed on to the United Nations Declaration on the Rights of Indigenous Peoples and committed to 'truth and reconciliation' with First Nations, these lofty goals have been and will continue to conflict with the exploitation that supports this county's natural resource-based export economy. Those goals also fly in the face of several persistent forms of neglect, in the form of persistent food insecurity, and insufficient access to health care. Dozens of First Nations communities in Canada have been under drinking water advisories, some for decades, due to the contamination of local water supplies.

Colonialism is not solely a rational enterprise of wealth accumulation. Because the means to that wealth entails stark ethical and moral transgressions of our relations with each other, with nonhumans and with the planet, as with patriarchy colonial projects require a system of justificatory beliefs: beliefs that justify entitlement, beliefs that justify genocide, beliefs that justify violence, as colonized peoples are deemed inhuman. As such, the capitalist ethos is deeply entangled and enabled by white supremacy.[70] Civilized, intelligent humans are transfigured into savage beasts; the perpetrators' acts of violence transformed into God's work. The civilized become re-written as the savage; savagery becomes civilized.

In turn, the living earth must be transformed into dead matter, by invoking human exceptionalism—the fallacious belief that humans are separate from, superior to, and ultimately capable of controlling, nature. This includes the separation of (White male) humans from nature, and the relegation of numerous Others—women, non-white, and Indigenous to the category of nature, and the subjugation of all deemed 'Nature' to Control. Land, resources, and peoples become objects subdivided into those which can serve the production of profit and wealth, and those which do not and are therefore expendable.

Just as with the management of emotions that allows capitalism to persist, this perceptual separation from nature inhibits our emotional responses to nature's degradation, transforming a 1000-year-old redwood tree into so much timber, the multi-million-year-old Appalachian mountains into stores of minerals, and the mighty Yangtze River into pent-up electricity waiting to be released to power new factories to make stuff, so much stuff. In the words of Amitav Ghosh,[71] upon achieving the imperialist project, the earth was no longer a subject of wonder, but of contempt. Indigenous peoples in turn are not only dispossessed of their land, their sources of livelihood, their rights to cultural expression, and their knowledge; they are also dispossessed of their very worldviews; worldviews which, although richly diverse, tend to share a belief that agency is not the restrictive privilege of humans alone, and thus seek to protect the sacred relationships between humans, non-humans and the places they inhabit.[72]

Emotional Consequences of Colonialism

This assertion of human superiority over and separation from nature harms us all, given the integral role our connections with nature have to human wellbeing. While we settlers are further removed, indoctrinated into human exceptionalism, the intimate relations with land, water, air, and nonhumans depicted by Indigenous elders are not restricted to Indigenous peoples alone. Psychological research offers strong support for this claim—opportunities to immerse ourselves in natural settings induce positive emotions, and the separation, or experience with sites of destruction, causes emotional harm.

For Indigenous peoples, this emotional harm is far more acute, as they never relinquished their ecological worldviews, and their attachment to land. This toll includes deep grief for the destruction of nonhuman relatives. But it also induces shame, for being prevented from fulfilling their cultural responsibilities, their prescribed duties to protect the land, water, and the planet's inhabitants, and provide food for their communities.[73] To complete the annihilation of Indigenous integrity, youth who are taught to be ashamed of their indigeneity turn away from traditional ways. As Kari Norgaard[74] illustrates in her work with the Karuk in Northern California, social interactions and cultural celebrations are intimately interlinked with local ecosystems. Take away the ecosystems and many of those essential practices to community vitality are lost. As one community member in her study lamented of the losses to younger generations,[75] 'you got to have fish to teach them how to fish.' For peoples for whom relations to land and nonhuman species are integral to identities, the severing of those connections as a result of ecological catastrophe is its own trauma. Many Indigenous peoples understand that the land and people must take care of each other, both are agents, both are animated, both are in need of care.[76]

But the emotional consequences of colonialism for Indigenous peoples does not end here. Their relocation and elimination of the means for subsistence and cultural vitality relegate colonized peoples to a position of dependency, of disablement, of being 'Othered' in one's homeland. This subjugation carries with it a deep emotional toll, but due to the violent character of these historical processes, this toll is compounded further by trauma.[77] Past traumas leave scars that are passed on to descendants, scars that are never quite given enough nourishment to heal. Instead, the wounds become re-inflicted, as when Indigenous women and girls are murdered, when the unmarked burial sites of neglected children attending residential schools are unearthed, when Indigenous peoples are targeted by police, or, they simply do not have access to clean drinking water. Merely being a bystander, as I am, is painful. But bystanders have the luxury of turning away, perhaps even going so far as to refuse to accept that we are capable of, and complicit in, such atrocities as have been committed over the past 600 years. We see such a rejection today, in repeated acts of discrimination, abuse, and neglect, in sceptical responses to news reports of thousands of children buried on the soils surrounding residential schools across Canada. The toll on Indigenous peoples is evident, in the form of drug and alcohol abuse, domestic violence, and suicide.

Closing off Our Discussion of Institutions and Structures

Capitalism and its bedfellows are pathologies that thread their way through our institutions, crippling our sense of autonomy, and our ability to build relationships based on trust and empathy, relationships that are crucial to our survival. They all, in different ways, impose mental stress on each of us, albeit far more so for marginalized peoples. When we are under stress for extended periods of time, several things occur, with consequences for both personal health and social wellbeing. According to neuroscientist Robert Sapolsky,[78] stress heightens the engagement of our hind brain to the detriment of our deliberative capacities. This in turn compromises empathy by reducing our ability to assess the emotions of others, and heightens our inclination to make a fear association during regular social interactions. Stress also compromises our ability to store and recall memories accurately, and increases the chances that we act impulsively or aggressively, sometimes just to let off the pent-up steam. All of which only serves to further increase stress, and none of which bodes well for our capacity to engage in cooperative responses to collective problems.

The more we conform to the institutions that embody these social structures, the more those structures are reproduced. However, institutional conformity is not guaranteed; it must be continuously and actively produced, and herein lay the possibilities for social change.[79] These structures have not yet completely smothered human agency, nor our innate drive for relationships and autonomy. Indigenous peoples the world over confront loss every day. Losses of loved ones, autonomy, capability, identity, family, and community are met with deep grief, and of course rage at the perpetrators—the settlers and their states—for their slow violence upon Indigenous land, peoples, and cosmologies. And yet, resistance to colonialism by Indigenous peoples persists; indeed, it has been growing, with #landback and Water is Life the new discursive frames of environmental politics today.[80] As noted by Bacon[81], rather than leading to surrender, despair has been driving active resistance to settler-colonial disruptions of Indigenous eco-social relations.

Rejection of patriarchy is also mounting today, in movements that link patriarchy to capitalism as well as colonialism. Schild[82] discusses the massive women's mobilizations taking shape in Latin America, giving voice to feminist principles that link demands for fair working conditions for women and eradication of sexual harassment to broader anti-capitalist struggles, and denouncement of the violent colonial invasion of territories, displacement of peoples, and ecological devastation in the name of resource-based development. Schild observes a new feminism emerging on the streets of Latin America that weaves together women's inequity, ecological destruction, and colonialism. Such connections are being drawn not solely in Latin America. Iraq's October Revolution of 2019 was, as much as anything, a confrontation not just with economic precarity but with that country's brutal patriarchy.[83]

Collective resistance creates space for the expression of emotions, and their transformation through sharing. Although social movement scholars have commented at length on the positive emotional appeal of social movement engagement, collective action also creates constructive spaces for the restorative expression of despair, grief and anger, as with Puerto Rican women mobilized to express their rage, or *coraje*, after the island was brought to its knees by a pair of climate-attributed hurricanes that made landfall in 2017, and then its peoples were summarily dismissed with the tossing of a roll of paper towels by then-President Donald Trump.[84] Resistance emerges in other forms as well, as with the engagement of the Chhara peoples of India— who have little reason for hope—in Budhan Theatre, a space Dia da Costa[85] describes as 'pedagogical spaces of compassion and critical optimism that performatively reconstructs a stigmatized 'criminal' community into a creative community.' And the revitalization of Black and Indigenous foodways, from South Dakota to Kerala.[86] These collective actions generate new alliances that cultivate efficacy and pride. The matrices of intersectional domination described earlier can define social alignments that spawn matrices of resistance, as awareness of common forces of oppression bring disparate groups together.[87].

What does all of this have to do with climate change politics? We turn to this question in the next chapters.

Notes

1 Bericat, "The Sociology of Emotions: Four Decades of Progress."
2 Bericat.
3 Meyer and Rowan, "Institutionalized Organizations: Formal Structure as Myth and Ceremony"; Bericat, "The Sociology of Emotions: Four Decades of Progress"; Bar-Tal, Halperin, and De Rivera, "Collective Emotions in Conflict Situations."
4 Voronov and Weber, "The Heart of Institutions."
5 Hochschild, *The Managed Heart: Commercialization of Human Feeling*.
6 Moon, "Powerful Emotions: Symbolic Power and the (Productive and Punitive) Force of Collective Feeling."
7 Creed et al., "Swimming in a Sea of Shame."
8 Creed et al.
9 Voronov and Weber, "The Heart of Institutions."
10 Pain, "Globalized Fear?"
11 Barbalet, *Emotion, Social Theory, and Social Structure*; Kemper, *A Social Interactional Theory of Emotions*.
12 Turner, *On Human Nature: The Biology and Sociology of What Made Us Human*.
13 Turner.
14 Rollert, "The Wages of Intimate and Anonymous Capitalism."
15 Moore, *Capitalism in the Web of Life*. P. 601.
16 Robinson, *On Racial Capitalism, Black Internationalism, and Cultures of Resistance*.
17 Collins, "Black Feminist Epistemology."
18 Liévanos, "Air-Toxic Clusters Revisited."
19 Bauman, *Liquid Modernity*.
20 Habermas, *Toward a Rational Society*.
21 Cited in Ferrarese, "Precarity of Work, Precarity of Moral Dispositions: Concern for Others in the Era of 'Emotional' Capitalism."
22 Pavletich, "Emotions, Experience, and Social Control in the Twentieth Century."
23 Weyher, "Re-Reading Sociology via the Emotions." P. 356.
24 Lutz, "Emotions and Feminist Theories."
25 Piketty, *Capital in the Twenty-First Century*.
26 Bauman, *Liquid Fear*.
27 Fineman and Sturdy, "The Emotions of Control: A Qualitative Exploration of Environmental Regulation."
28 Turner, "Self, Emotions, and Extreme Violence." PP. 520–521.
29 Martin, "On the Persistence of Fear in Late Capitalism."
30 Giddens, *The Consequences of Modernity*; Bauman, *Liquid Modernity*.
31 Beck, *Risk Society*.
32 Beck.
33 Ahmed, "Collective Feelings: Or, the Impressions Left by Others." P. 119.
34 Pain, "Globalized Fear?"
35 Rico, Guinjoan, and Anduiza, "The Emotional Underpinnings of Populism"; Salmela and Von Scheve, "Emotional Roots of Right-Wing Political Populism."
36 Lemke, "3. The Risks of Security."
37 Ahmed, "Collective Feelings: Or, the Impressions Left by Others."

38 Cited in Ferrarese, "Precarity of Work, Precarity of Moral Dispositions: Concern for Others in the Era of 'Emotional' Capitalism."
39 Iyengar, Sood, and Lelkes, "Affect, Not Ideology"; Stewart, Plotkin, and McCarty, "Inequality, Identity, and Partisanship: How Redistribution Can Stem the Tide of Mass Polarization."
40 Törnberg, "How Digital Media Drive Affective Polarization through Partisan Sorting."
41 Guthridge et al., "A Critical Review of Interdisciplinary Perspectives on the Paradox of Prosocial Compared to Antisocial Manifestations of Empathy"; Törnberg, "How Digital Media Drive Affective Polarization through Partisan Sorting."
42 Törnberg, "How Digital Media Drive Affective Polarization through Partisan Sorting."
43 Martin, "On the Persistence of Fear in Late Capitalism"; Salmela and Von Scheve, "Emotional Roots of Right-Wing Political Populism."
44 Salmela and Von Scheve, "Emotional Roots of Right-Wing Political Populism."
45 Webster and Albertson, "Emotion and Politics."
46 Webster, *American Rage*.
47 Simas, Clifford, and Kirkland, "How Empathic Concern Fuels Political Polarization."
48 Salmela and Von Scheve, "Emotional Roots of Right-Wing Political Populism."
49 Folbre and Nelson, "For Love or Money—Or Both?"
50 Ferguson et al., "Introduction." P. 31.
51 Walby, *Theorizing Patriarchy*.
52 Gilligan and Snider, *Why Does Patriarchy Persist?*
53 Rached, Hankir, and Zaman, "Emotional Abuse in Women and Girls Mediated by Patriarchal Upbringing and Its Impact on Sexism and Mental Health."
54 Seremetakis, *The Last Word*.
55 Pease, "The Politics of Gendered Emotions."
56 Rutherford, *MEN'S SILENCES Predicaments in Masculinity*.
57 Pulé, Hultman, and Wägstrom, "Discussions at the Table."
58 Brescoll and Uhlmann, "Can an Angry Woman Get Ahead?"
59 McGregor, "Conceptualising Male Violence against Female Partners"; Pease, "The Politics of Gendered Emotions"; Walton, Coyle, and Lyons, "Death and Football."
60 Lutz, "Emotions and Feminist Theories."
61 Rached, Hankir, and Zaman, "Emotional Abuse in Women and Girls Mediated by Patriarchal Upbringing and Its Impact on Sexism and Mental Health."
62 Al-Hassani, "Iraq's October Revolution." P. 109.
63 Gilligan and Snider, *Why Does Patriarchy Persist?* P. 50.
64 Mellor, *Breaking the Boundaries*. P. 155.
65 Oksala, "Feminism, Capitalism, and Ecology." P. 220.
66 Alfred and Corntassel, "Being Indigenous."
67 Bacon, "Settler Colonialism as Eco-Social Structure and the Production of Colonial Ecological Violence."
68 Bacon. P. 59.
69 Schild, "Feminisms, the Environment and Capitalism: On the Necessary Ecological Dimension of a Critical Latin American Feminism."
70 Holleman, "De-Naturalizing Ecological Disaster."
71 Ghosh, *The Nutmeg's Curse*.
72 Watts, "Indigenous Place-Thought & Agency amongst Humans and Non-Humans (First Woman and Sky Woman Go on a European World Tour!)."
73 Norgaard and Reed, "Emotional Impacts of Environmental Decline."
74 Norgaard, *Salmon and Acorns Feed Our People*.

75 Norgaard. P. 207.
76 Kimmerer, *Braiding Sweetgrass*.
77 Brave Heart et al., "Wicasa Was'aka."
78 Sapolsky, *Behave: The Biology of Humans at Our Best and Worst*.
79 Moon, "Powerful Emotions: Symbolic Power and the (Productive and Punitive) Force of Collective Feeling."
80 Richez et al., "Unpacking the Political Effects of Social Movements With a Strong Digital Component."
81 Bacon, "Settler Colonialism as Eco-Social Structure and the Production of Colonial Ecological Violence."
82 Schild, "Feminisms, the Environment and Capitalism: On the Necessary Ecological Dimension of a Critical Latin American Feminism."
83 Al-Hassani, "Iraq's October Revolution."
84 LeBrón, "Policing Coraje in the Colony: Toward a Decolonial Feminist Politics of Rage in Puerto Rico."
85 Da Costa, "Cruel Pessimism and Waiting for Belonging." P. 3.
86 Kozhisseri, "'Valli' at the Border: Adivasi Women de-Link from Settler Colonialism Paving Reenchantment of the Forest Commons."
87 LeQuesne, "Petro-Hegemony and the Matrix of Resistance."

6 Inaction Pathways
On Why We Don't Do the Things We Don't Do

Inaction in the face of a global climate emergency, as maladaptive as it is, is really no surprise. Some would say it is even logical. Mancur Olson[1] would certainly say so; he argued convincingly that engaging in collective action to achieve benefits for the common good makes no sense, at least for rational, self-interested actors, since the individual inevitably invests more in effort than the share of reward they are likely to receive. Fortunately, Olson and many of his contemporaries vastly over-estimated just how many of us can be described as purely rational self-interested actors. Nonetheless, for many reasons, reasons that are different for different people, inaction continues to rule the day.

Let's face it, doing nothing, or at least, doing as little as possible, is easy. I will risk falling out of my chair trying to reach for that book I want, sitting on the shelf that is so close but not quite close enough, rather than just standing up to get it. Sticking to one's beliefs, comforting as they are, is also easy, even in the face of conflicting evidence. Following leaders, rather than leading, so easy peasy. Being told what's right and what's wrong, rather than figuring it out for ourselves, also very easy. Even despair is an easy if not exactly comforting place to sit, certainly easier than fighting a behemoth against painfully slim odds. The many reasons for the high levels of climate inaction we observe today can be attributed to the nature of the trigger (information), the unique nature of each of us, and the nature of the social systems within which both are embedded. In combination, these three elements align to generate different emotional pathways, many of which favour inaction, including apathy, denial, withdrawal, and being stuck. Each of these will be explored in detail below, but first, let's focus on triggers, actors, and systems.

The Trigger

Weather is something we all experience directly, and viscerally. And it is something in our lives that routinely has our attention, if its prominence in our daily chit chat, in news media, and in books and movies, is any indication. We check the weather daily to determine what to wear, and whether the soccer game is likely to be held or not, and are primed to watch for the storm on the horizon. If only the climate emergency were 'just' a storm on the horizon!

But climate is not weather, and to recognize a change in climate, to be able to attribute that change to some cause, and to realize its implications, requires interpreters. Thus, climate emergency triggers are inevitably indirect, filtered through information providers. To begin with, there is much room for misreading of the trigger, based on the nature of the information itself. Global warming is, on the surface, not that complicated to understand—we have covered our atmosphere in a blanket, trapping heat underneath. And heat not only makes land, oceans, and bodies warmer, it is also crazy-making for our planetary systems—weather, ocean currents, disease vectors. This is very, very bad news for our social and economic systems, which have come into their own in a historic period of relative climatic stability. So global warming as a phenomenon is not hard to understand, but the *study* of climate involves some pretty sophisticated scientific research, tools, and language, and let's face it, climate scientists often communicate in terms that are enormously abstract and complex for the untrained.[2] It is any wonder many fail to be alarmed about global average temperatures (two degrees *really* doesn't sound like a big deal to me!), parts per million (that's so *tiny*!), and the year 2080 (wait, when? I'm trying to make it to the end of the month!), and always with uncertainty bars, which are just part of the language of science, but are not received well by the rest of us, and certainly not policymakers (okayyyy, so you're saying you don't actually know for sure?).

More to the point, scientific messages tend to be in the language of, well, science[3]. A scientifically trained eye may find a particular temperature graph or departure from the mean alarming, but non-scientists are not likely to react in the same way. We respond to stories, not statistics and graphs. Walter Fisher[4] describes we humans as 'homo narrans'; we have evolved to communicate through narratives.[5] And stories grab our attention, or not, based on the degree to which 'a plot activates the story receiver's imagination through an empathic connection with the characters.'[6] Who are the characters in this story? What's the plot? And how does this story jibe with my understanding of how the world works, and what's important? Time to change the channel.

With the spectacular rise in frequency of truly unprecedented disasters in recent years, many have speculated that, at long last, a great awakening must inevitably follow: direct experience with exactly the kinds of impacts projected by climate scientists—floods, heat waves, and fires—must surely generate alarm about our climate emergency, right? Alas, even the floodwaters covering your neighbourhood can be subject to interpretation, and several studies suggest that direct experience with a climate-attributed disaster has only a modest effect, if any, on concern about climate change.[7] This is not too surprising, since the attribution of such events to global warming still requires the intervention of an information provider, and that's difficult, since floods are hardly a new phenomenon. To further dampen the trigger, until recently journalists have been very wary of discussing climate change in their coverage of extreme events.[8]

Some studies do indicate increases in concern levels subsequent to direct disaster experience, but this effect is far more likely among survivors who already agreed with climate science before the event.[9] This is particularly the case for survivors holding liberal political views, although some studies do show that climate concern can increase even among survivors with conservative political leanings.[10] Sugerman and their colleagues conducted a rare meta-analysis of studies and found a 1°C increase in temperature increases worry about climate change by 1.2%;[11] statistically significant yes, but teeny really.

We not only assess the information package itself, we also inevitably assess the information provider, and here there is yet more space for missing the trigger. Michael Lynch[12] discusses how we tend to rely upon proxies to assess the credibility of claimsmakers. So, as with any news or gossip, we rapidly determine how much credence to give to the claim; too rapidly in many cases, relying on heuristics rather than deliberation. Am I already familiar with this particular claimsmaker? If not, what are their claims to authority? Do they represent an institution I trust? And most consequentially, but least acknowledged, do they look and talk like me?

Scrutiny of claimsmakers is absolutely important, because certain climate science claims can be quite lucrative to certain interests, and due to its complex and abstract character, climate science can be readily subjected to interpretation—many interpretations in fact. Here is where the agency of the information providers comes in, many of whom do indeed have particular interests. On the one hand are claimsmakers backed by a well-resourced climate denial machine, on the other side are advocates seeking to motivate pro-climate action, and stuck in the middle are journalists who were trained to present a 'balanced' story, while avoiding offence to their state or corporate masters.

Unfortunately, while climate denial messaging has been found to be enormously effective at supporting inaction, the reverse is not necessarily true. There has been a concerted effort to evaluate the efficacy of climate messaging for supporting pro-climate action, and on the whole this research record indicates that such messaging has rather minimal, and temporary effects, if any.[13] One clear finding from this work, in alignment with Fisher's theory, is that messages evoking emotion are far more effective at getting attention than messages that do not.[14] But the findings regarding emotion-laden messaging are messy. Messages evoking negative emotions appear to be more effective at eliciting reactions,[15] but while messages seeking to convey the scale of the problem should and do motivate pro-climate action, they also can induce anxiety, which can contribute to a sense of helplessness and despair.[16] On the other hand, research participants tend to respond more positively to hopeful images, of climate solutions for example, but these messages can be even less likely to motivate action than are messages evoking negative emotions like fear and anxiety.[17]

On the whole, while there is strong agreement that transforming the way we talk about climate change is required to generate widespread pro-climate action,[18] media messaging may be necessary but by no means sufficient on its own. As Joanna Wolf and Susanne Moser[19] state simply, people are not blank slates. For this reason, Susanne Moser and Lisa Dilling[20] argue that face-to-face communication will inevitably be more persuasive for several reasons:

> first, it is more personal; second, nonverbal cues can allow the communicator to gauge how the information is being received in real time and respond accordingly; direct communication also allows for dialogue to emerge; and finally, the trust between individuals participating in a two-way exchange goes a long way toward engaging and convincing someone.

The Actor

So, yes, we are all unique, and we are conscious beings each of whom is capable of deliberating upon and choosing courses of action according to our unique sets of personal values. But, as noted in Chapter 4, we also share in common 99.6% of our DNA. That has to generate certain commonalities. We can all only go for so long before we have something to eat, for example, and we tend to metabolize our food in much the same way. Our commonalities extend into the neurological terrain as well. A few commonalities stand out. First, humans share a propensity to acquire and conserve energy with all other living beings on earth, for the simple reason that to do so is to survive. Even when survival is not at stake, however, that inclination to conserve energy remains. Energy conservation also means conserving our mental effort, both cognitive and emotional. We also talked about this in Chapter 4. Why perform longhand division when there is a calculator in your pocket? Why stop and assist someone in need when it is so much easier to look the other way, thereby making it to my meeting on time? Why learn new ways of doing things, when the way you've done them for years works just fine? This is not to say that we in all cases abide by this metabolic programming, but in order to choose to expend energy when there are other, less costly avenues available to us, we must have a good reason.

This energy efficiency hard-wiring expresses itself in many ways, but one in particular is a process called psychological distance, a phenomenon that many researchers have associated with low levels of concern about climate change, which is often perceived as something that is only relevant to faraway places (the Arctic say), or in the distant future. If you want to dig deeper, or just want a fancier academic term, read up on what psychologists refer to as Construal Level Theory. The basic idea is this: proximity, or closeness to a threat or opportunity, reflects the degree of emotional stimulation engendered. According to psychologists Jiayi Luo and Rongjun Yu,[21] our emotionality has evolved to assist us with rapid, intuitive responses to immediate threats, and immediate opportunities. In contrast, we are slower to respond to distal threats and opportunities. This applies to spatial distance, temporal distance, and social distance. As a result, we are still perfectly capable of attending to more 'distant' phenomena, but doing so requires overriding our first inclination to ignore them, and as a result is more likely to require cognitive effort. 'Distance' is also inevitably subjective, being in reference not solely to a threat, like a hurricane, but the implications of that threat to the things we value. Threats that appear to be far away, in time, in space, or otherwise remote *from the things and people we care about* will generally get less of our personal attention than otherwise. On the other hand, feeling strong emotional attachments towards something under threat that may be far away can reduce our psychological distance from an object or event, describing a two-way relationship.[22] Despite the fact that I have not lived in California for several decades, for example, news about negative events in California still affect me on an emotional level far more than news about, say, Manitoba, which is roughly the same distance away. The reference point for the threat also matters. In highly individualized Western cultures, the material wellbeing of oneself and family may well be the primary reference point from which threats are evaluated, but this is not necessarily the case in cultural settings in which communities, landscapes, and/or nonhuman beings are viewed as equally important.

Second, and relatedly, as discussed in Chapter 4, positive emotions feel good, are energizing, and we all would much prefer to experience them than their opposite. Negative emotions feel bad, are draining, and we don't like them (and this fact is the bread and butter of the pharmaceutical industry, legal and otherwise). Global warming, unfortunately, does not invoke many positive emotions. This does not mean that we all inevitably heed the hedonic goal of feeling good, but it does mean that intentionally approaching situations that evoke negative emotions is personally costly. Situations that evoke positive emotions include not just things like new clothes (however temporarily); among other things, reinforcement of social belonging, and validation of personal identity also feel really, really good. Reinforcement of self-determination, and sense of personal control, as indicated in, for example, goal achievement, also feels great. Situations that evoke negative emotions include some obvious things like fear and sadness, but also threats to belonging, as with shame and guilt. For someone who is concerned about the climate emergency but is immersed in a social group in which mentioning such fears is enough to lead to heated conflicts, and even ostracism, this matters a whole lot. Threats to self-determination, as with high levels of uncertainty and feelings of powerlessness that are often evoked by the climate emergency, also elicit negative emotions. And finally, because of our empathy, we also experience negative emotions when observing subjects we care about under duress, and we tend to react strongly to perceived unfair or unjust outcomes, or outcomes that otherwise conflict with one's moral compass. These emotional responses, as we will see in the next chapter, can motivate pro-climate action, unlike the other negative emotions discussed here.

Third, despite where I began this section, it bears repeating that we are not all cut from the same cloth. This means that while we all share both of these two inherited processes, they will manifest under very different circumstances for different people. What constitutes a trigger of negative emotions like shame, or what we perceive is under threat or too distant to worry about, depends upon one's intersectional positionality, but also their cultural context, and personal values. There is also, simply, much room for variation, based on interactions between complex genomes and environments. Some of us, for example, hold hedonistic values for pleasure and personal gratification to be paramount; others have much higher regard for the welfare of other beings, and the community as a whole. Still others care deeply and feel responsible for the natural world.[23] Some of us are more inclined to respect authority, others more concerned about equity. These inclinations can encourage gravitation towards certain in-groups that offer reinforcement, leading to, among other things, a form of 'moral tribalism' within social and political groupings, priming individuals for very different reactions to the same information and experiences.

The System

There is much that can be said about how contemporary social systems intervene in our exposure to and evaluation of climate emergency triggers, particularly the polarized, racialized, and gendered nature of modern western social structures. I covered a lot of this terrain in Chapter 5 that I won't repeat here. It bears highlighting one thing though. First and possibly foremost, in our global capitalist system, information is a commodity. Yes, information can have tremendous use value to the holder of that information, and information access is an advantage that not everyone enjoys. But information is more than this—it is bought and sold by information traders who care not a whit for its use value—including its accuracy or its consequentiality. The only interest of information traders is the profit-making potential embodied in its exchange value. And information is very profitable indeed—the largest corporations today are all in the information business—to the extent that it grabs attention, attention that can then be sold to advertisers.

Anytime we convert an essential need into a commodity, the potential for negative social consequences emerges. We have many other things that fall into this category, like food, medicine, and housing. And this is one reason why these industrial sectors tend to involve higher levels of government intervention than other sectors of the economy—such interventions are attempts to protect consumers from the deleterious consequences of commodifying essential needs, to widely varying success. Information is also an essential need, but one for which there has been far less government intervention to protect the consumer in most countries. As such, since the exchange value to information traders of accurate information regarding the climate emergency may be limited, it can often be outweighed by the exchange value of, say, heated scientific 'debates,' sceptics' claims that we don't have to change our lives after all, or better yet, juicy conspiracy theories. And while there certainly are good reasons to critique the increasingly corporatized nature of our mainstream news industry, established media institutions do at least ascribe to journalistic norms of accuracy and objectivity. No such norms, however, are imposed upon the cornucopia of alternative and social media enterprises.

Four Pathways to Inaction

Considering the messy realities I've just described, there are any number of emotional responses that could lead one towards a position of inaction in response to the climate emergency. And while creating categories always simplifies the fluid nature of complex systems, I think *most* of those pathways can be grouped into four general pathways: apathy, denial, withdrawal, and stuck.

Apathy

This bucket captures all those who accept that climate change is happening, but are not particularly emotionally triggered or concerned, and do not feel motivated to take any action in response. This does *not* describe individuals who are concerned but unable to attend to global warming due to their limited capabilities, whom my colleagues and I refer to as inerts, or simply 'stuck.' Rather, apathy, simply speaking, refers to a lack of concern despite awareness, and can be observed across all income levels. In the aggregate, apathy can culminate into what sociologist Anthony Giddens referred to as the Giddens' Paradox,[24] referring to the seemingly very likely probability that by the time we give the climate emergency our full attention, by the time we are slapped with enough blatant warnings, it will be too late to avert it, and the best we can do is try to pick up the pieces after the storm, only to have another storm waiting to descend. Even though awareness and concern levels are rising, several studies continue to show that most people in Global North countries—where much of our empirical record is concentrated—do not feel *personally* threatened by climate change.[25] In other words, for many the climate emergency remains psychologically distant, and studies confirm that perceived psychological distance from climate change is associated with low levels of concern,[26] and consequently high levels of inaction.[27] Most of us also have an inclination towards wishful thinking, or what psychologists call optimism bias: the tendency to over-estimate the likelihood of positive outcomes. Without it, there would be no gambling industry. Optimism bias doesn't just support gambling, though. If I wasn't a wishful thinker, I would never get in a car again, considering the decent odds of an accident. A bit of wishful thinking helps us get through the day without becoming completely overwhelmed by fear and trepidation about the dangers around us. It also inclines us to under-estimate worst-case scenarios, however, and failing to recognize the potential for such worst cases leaves us woefully unprepared for them when they do occur. Optimism bias includes technological optimism, confidence in expertise and authorities, or just confidence in one's own sense of control, a phenomenon named the 'resilience paradox.'[28] In a study of Chinese youth, for example, Lei Shao and Guoliang Yu[29] found that those with the highest reported levels of optimism (measured as perceived resilience) were least likely to engage in pro-environmental behaviours. This translates into a lack of preparedness, and also a lack of any attention to the problem whatsoever. *Why worry?*

While the term apathy implies carelessness, a not-caring about something, Ezra Markowitz and Azim Shariff[30] argue that even for those who view the wellbeing of others as a priority, 'the human moral judgement system is not well equipped to identify climate change—a complex, large-scale and unintentionally caused phenomenon—as an important moral imperative.' Consequently, I use this label a bit more broadly, to capture all those who simply have other priorities demanding their attention. Or perhaps the actor is an eternal optimist, or just has a lack of mental space to reflect upon this enormous, complex threat that feels so far away. Perhaps apathy is in part the result of the narrowing of our emotional gaze under capitalism, in which individualization, materialism, short-term thinking, and human exceptionalism are perfectly normal—even rewarded—ways to make one's way through life.

Many researchers speculate that this non-responsiveness is linked to the way the information was presented, failing to elicit an appropriate trigger warning. The subject of climate change remains insufficiently covered in schools, and climate change information is often presented in ways that limit making associations between global warming and one's personal wellbeing and the things one values. As already discussed earlier, the topic is complex, and often the impacts one hears about refer to happenings far away in space and time. That information is also presented in terms that are meaningless to non-scientists, in the form of, for example, increases in carbon dioxide concentrations. As stated by Sheila Jasanoff,[31] 'when it comes to nature, human societies seem to demand not only objectively claimed matters of fact but also subjectively appreciated facts that matter,' and CO_2 molecules just don't seem like something that should matter outside of chemistry class. And then there is a whole lot of conflicting information available, leading many to throw their hands in the air.

To add to the insulating effects of overly complex and abstract or conflicting information messages, many of us are insulated in other ways. While access to interpretive information is essential, exposure to the elements, and opportunities to read changes in landscapes do matter, particularly by personalizing the impacts of global warming. Many of us live in ways that are highly disconnected from our environment, however, living in modern, urban landscapes covered in cement and steel, and the daily routines of many involve movement between indoor spaces, with spending no time at all outside in a typical day.[32] One's level of privilege, furthermore, directly translates into one's ability to insulate from personal experiences of global warming's negative impacts, in the form of air conditioning, for example.[33] Extreme heat days really don't feel like an emergency for those who can simply remain within air-conditioned homes, cars, and offices.

We can become desensitized to climate change information in other ways too. Psychologists have pointed out that our empathy primes us to react to identifiable victims: the abused next-door neighbour, the lone, starving polar bear mom and her cubs caught on camera, or the clearcut on the ridge.

We are far less likely to have the same reaction to information regarding threats described in statistical, or other aggregate forms.[34] Markowitz and colleagues refer to this as 'compassion fade.'[35] We also just might be experiencing cognitive and emotional overload. After our emotional attention is consumed by grades, job security, housing bubbles, medical bills, rising food prices, police violence, pandemics, the fragile state of democracy, war, worrying about whether it's time to consider getting your parents into a nursing home, or the heroic job—especially for single parents—of getting your kids to school with their teeth brushed, lunches packed, and permission slips signed, every morning, five days a week, there just isn't much emotional energy left.

Denial

Alright, let's move on to the big one: denial. Not big in size, since denial is thankfully on the decline and now represents a minority of populations, but it remains big in political effect. First, let's clarify just who we are talking about. There is a small, concentrated minority of vested interests who have orchestrated an enormously successful discursive campaign to sow doubt about the validity of climate science, broadly speaking. Denial discourse represents a spectrum, from 'hard' to 'soft' denial claims, with hard denial representing an absolute rejection of the scientific finding that global warming is caused by humans, or even, for a few, that global warming is happening at all. Soft denial claims include a variety of positions that acknowledge that global warming is happening and is at least partly caused by humans, while rejecting that we really should be bothered by this fact—a few degrees of warmth can be pleasant after all, and plants love CO_2!—and we most certainly should not be investing our money and energy into changing the way we do things, least of all phasing out fossil fuels, the backbone of our economies. Extensive research shows that climate denial is by no means the outcome of some sort of rational, cognitive evaluation of evidence on the part of individuals; if it was, denial would be distributed randomly throughout the population, perhaps governed only by things like literacy, information access, or education levels. To the contrary, these factors do not seem to make much difference, so let's unpack the factors that do.

There is an important and powerful body of research that focusses on the 'Denial Machine': the well-organized and highly effective disinformation campaign funded by right-wing think tanks and other vested interests. There is a wealth of research on the Denial Machine for those interested in learning more; if you want to dive in I would start by checking out the Climate Social Science Network. Here, however, I am far more interested in those who find the claims being made by the members of the Denial Machine so palatable. I believe this group includes individuals for whom the findings presented by climate science—in other words not climate change itself but climate change information—is a threat, a threat to lifestyles, to identities, and to belief systems, that is so overwhelming that alternative storylines are welcomed with a big sigh of relief.[36] And let's face it, who wouldn't jump for joy to learn that the climate emergency was just some giant hoax, and everything is FINE? I know I would. As Bruno Latour lamented in one of his last interviews, aren't we all climate deniers at some level? Robert Brulle and Kari Norgaard[37] describe climate change as a cultural trauma, particularly for those raised on Western diets of progress, materialism and individuality. This assertion is based in part on earlier work by Kari Norgaard,[38] evaluating the reactions to global warming of citizens in a small Norwegian town. Norgaard found that people avoid thinking about climate change in part because doing so raises fears of ontological insecurity and helplessness, and poses a threat to individual and collective senses of identity. And the guilt! That horrible feeling of responsibility for having contributed more than our fair share to a terrible problem. Denial of responsibility is a great way to alleviate the inevitable guilt that surfaces when we contemplate the vast differences in per capita emission levels.[39]

Norgaard's research subjects were anything but exceptional. Coming to terms with global warming inevitably leads one to come to terms with the layers of privilege and marginalization we each differentially experience. Settlers must come to terms with colonialism, patriarchs must come to terms with sexism, whites with systemic racism. The wealthy must come to terms with the gross inequities supported by modern neoliberal capitalism, and Western scientists, engineers and technophiles must accept that Nature will not be put under our thumb. It requires acknowledging that the many forms of justification and social order that have offered comfort and stability (for some at least) over the past century and a half are a facade. It is not climate change itself, but rather this second layer of truths that keeps many of us hiding under the covers of denial, only too ready to grasp onto fictions rather than truths. The knowledge of climate change, rather than its anticipated impacts, constitutes the threat; a threat to vested interests of course, but also to identities, values and worldviews, and perhaps most intimately, western lifestyles. Fossil fuels and industrial agriculture are deeply entwined in our households, economies, and politics, empowering vested interests while also dampening threat appraisals among the rest of us in the western middle classes, as life without them seems unimaginable. In an intriguing new study, researchers were able to corroborate these arguments, showing that individuals who attribute global warming to natural causes experience lower anxiety and fear in comparison to those who attribute global warming to anthropogenic causes.[40]

All this adds up to plenty of support for climate denial. But we aren't all *equally* prone to denial. In fact, climate concern has been steadily on the rise. There are nonetheless plenty of people who still fall into this group, particularly in Anglophone countries like Canada, the U.S., the U.K., and Australia. Climate obstruction researchers have attributed this regional clustering to the fact that the Denial Machine has been most active in these countries. Beneath these regional disparities, we have a pretty good understanding of just who is most prone to denial within these regions, based on decades of work in environmental sociology and environmental psychology, and there are three classifications that come to the fore: conservative, white, men. A tendency for white men to be less concerned about all kinds of environmental and technological risks has been well-documented in research dating back 30 years. More recent work, spearheaded by Aaron McCright and Riley Dunlap,[41] has really taken this linkage to new heights, beginning with research that brings to the fore political ideology: not *all* or even a majority of white men are necessarily so dismissive of environmental, and now climate, concern. Rather, it is that subgroup of white men who lean conservative who are most likely to fall into this category. And while conservative women express higher levels of concern overall than their male counterparts, conservative women nonetheless express less concern than liberal-leaning women. On the other hand, political ideology appears to have a much stronger influence over climate change attitudes among white people than among non-whites.[42] This lack of concern then readily transpires into rejection of the warnings heralded by environmental and climate scientists and advocates, ergo climate denial.

This work signals something that we have been discussing throughout this book: the identities of each one of us are deeply entangled with the groups to which we belong, and these identities serve as information filters. Even a flood can be interpreted differently by different survivors: Republicans reported significantly less neighbourhood damage than non-Republicans after a flooding event in Colorado,[43] regardless of their proximity to the damage, and Larry Hamilton and colleagues[44] found that Democrats were more inclined to acknowledge an observed increase in regional flooding than Republicans.

The emotionality associated with climate denial, and its relationship with identity does not stop here, however. The more attached we are to a particular group identity—say a Christian, a logger, or an American—the more we will be especially susceptible to embracing and defending the beliefs, perspectives, and values espoused by that group, particularly when disconfirming evidence comes from out-groups.[45] Flag-waving group adherents are also only too ready to intuit and then react to threat cues directed towards one's in-group, such as allegations of causing climate change, or claims that the lifestyles of one's in-group are no longer acceptable, things like driving trucks, eating meat, or taking winter holidays. Threats to social identity generate collective or group shame—in other words shame on the behalf of the group.[46] As stated by Beth Yeager,[47] 'collective shame ... is denied, dissociated from, and buried behind defenses such as denial, anger, blame, and conduct that appears shameless.' Add to this the echo chambers facilitated by our web-based social and information environment and we get heightened polarization, reinforced through emotional attachment to one's in-group, and hatred, fear, and distrust of out-groups.[48] White conservatives have been mobilized around the belief that a liberal elite has sought to shame and stigmatize them.[49] Similarly, Robin Veldman and colleagues[50] found that a sense of marginalization and embattlement signaling a threatened social identity was at the centre of evangelical Christians' climate denial. And a strong attachment to a U.S./American identity has been associated with denying or justifying environmental harm.[51]

So, social identities, our in-groups, matter to our responses to the climate emergency, but only some identities seem to be associated with denial. Why would membership in these groups in particular lead to higher rates of climate denial? After all, many in these groupings are well-educated, have access to information, and for many, the resources to invest in transition. To answer that question, let's start with a closer look at the strongest indicator of the three, political orientation. Political orientation has been shown to have a stronger and more consistent effect on concern and its antithesis, denial, than any other demographic variable, including race and gender. Partisanship comes into play of course, across the political spectrum, particularly negative partisanship, or the inclination to oppose a policy solely because the opposing party supports it. Negative partisanship can even be a stronger motivator than positive partisanship, or allegiance to one's own party, as Adam Mayer and Keith Smith[52] found in their study of climate policy support in the U.S., Republicans who were highly motivated by opposition to Democrats were less likely to support climate policy than those Republicans who were mainly driven by support for their own party. We may all be susceptible to negative partisanship, and of course positive partisanship as well, but it is the leadership of liberal parties who express the greatest support for climate action, and those of conservative parties across many jurisdictions who are most likely to express a climate denial position. A conservative political orientation is also associated with a specific set of core values, some of which are directly threatened by the reality of human-caused global warming. Acknowledging, and by extension responding to global warming and environmental destruction conflicts with prescriptions for limited government, free market enterprise, economic growth, and support for capitalism generally.[53] In other words, conservatives generally seek to defend the current economic and governmental orders in their countries and resist change.

But wait, there is yet more to the link between conservatism and denial that I for one find fascinating, and it seeks to explain not only the strong correlation between conservative orientation and climate denial, but also *why* certain people lean into conservative ideologies in the first place, while others lean liberal. After all, unlike many other identities such as race and nationality, we are not *assigned* membership in political identity groups; they are chosen. And our selection of political orientation doesn't necessarily begin with an attraction to certain leaders, or specific policy positions. Our identification with different political ideologies can come from a much deeper place, that place where we hold certain basic values, and two in particular that are linked to conservatism are a social dominance orientation, and relatedly, respect for authority, which can fairly readily slip into support for authoritarianism. In other words, conservatives are more inclined to look to their leaders to tell them what to think, believe, and do. This respect for authority can help explain the strong deference observed among Republicans to their leadership. Conservatives, more so than others, tend to support policies not based on the merits of the policy itself; rather, it is enough to have one's leader offer support for that policy. In other words, conservatives are more likely to show a strong deference to party leadership. Republicans, for example, are more likely to likely to express concern about, or belief in, climate change when their Republican representatives express concern.[54] In research conducted on the 2016 U.S. presidential election, Ulf Hahnel and colleagues[55] revealed a positive emotional shift towards the Republican Party, which was directly linked to reductions in belief in global warming.

Recent research[56] indicates that embracing a social dominance orientation and authoritarianism may be even stronger predictors of climate denial, and anti-environmental views generally, than political party affiliation. In other words, even though people who hold these values tend to gravitate to conservative political parties, there is nonetheless some variation in the level of ascription to these values among conservative party members, and the higher the ascription, the higher the likelihood for climate denial. The association between conservative party membership and climate denial, then, can be blamed partly on the rhetoric of conservative opinion leaders to be sure, but such rhetoric is not simply passively absorbed; the receiver has to buy it. So, the tendency towards climate denial among people with a conservative political orientation may have to do with the attraction to conservative parties of individuals espousing a social dominance orientation and authoritarianism, for some more than others.

Authoritarianism is pretty self-explanatory, but social dominance orientation deserves a bit more explanation. A social dominance orientation depicts ardent support for social hierarchies, viewing inequalities as a form of 'natural order.' If I have more money, influence, opportunities for promotion, or other successes than you have, it must be because I deserve it. And you don't. Certain types of people are simply naturally superior, and hence more deserving, than others, according to a social dominance orientation. No need for policies to correct inequities then, and they wouldn't work anyway, since they go against the natural order of things. This perception of the world as a sort of competitive jungle in which dominant and submissive players sort themselves out and those on top deserve their just rewards, is also extended to an ascription of human dominance over nature.[57] As should be no surprise, ascription to a social dominance orientation seems to be related to low levels of empathy, as well as a general desire for power and control in relationships.[58] Both authoritarianism and social dominance orientation also flourish in right-wing extremist movements, taking such sentiments to further heights, in the form of white supremacy and fascism. And right-wing extremist movements, which have been on a worryingly ascendant trajectory over the past decade, have adopted climate denial enthusiastically, which is breathing new life into hard denial discourse.

Now, let's talk about gender. Why would women tend to be more concerned, and men less so, about environmental and climate disruption? This isn't about biology; it is about social roles assigned on the basis of gender. Perhaps not surprisingly, much of the scholarship that has explored gender differences in environmental concern has tended to focus on women—asking what is it about women that causes them to care more about the environment? Much research has drawn attention to traditional divisions of labour in the family, in which women are more likely to be responsible for caring for children, and caring for pretty much everyone else in the family and community, even when they also participate in the workforce. This leads to greater sensitivity to the threats to family and community wellbeing posed by environmental and climate disruptions.

The work that has focused on why men are inclined to be *less* worried and more likely to reject environmental and climate science is smaller but even more interesting. To begin with, once again, it should be noted that we are primarily talking about *white* men here. While there are indeed gender dynamics that feature in other racial and ethnic groups, including many that are governed by patriarchy, the systemic colonialism discussed in Chapter 5 means that white men have enormous privilege in comparison to non-white men; and under that system, non-white men have been belittled and dehumanized along with women. In one of the earliest empirical studies revealing the extent to which white men stand out among all other groups in their lower concern levels, Flynn, Slovic, and Mertz[59] speculated simply that 'perhaps [white men] see less risk in the world because they create, manage, control and benefit from so much of it.' That is certainly true, but new works, by Martin Hultman and Paul Pulé,[60] and Cara Daggett,[61] and their colleagues, direct our gaze not to men as such, but to masculinity and the masculine social identities towards which men are taught to aspire. These can also take on culture- and race-specific flavours, but my graduate students and I have identified several elements that can be linked to climate denial, which appear to transcend many specific masculinist identities. First, under a masculinist ethos, concern for environmental and climate disruptions is feminized.

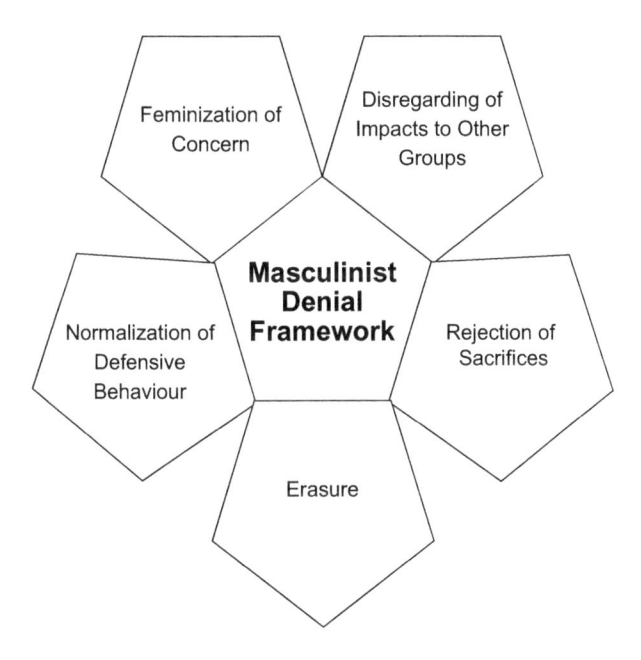

Figure 6.1 Masculinist-denial framework. Originally published in Davidson et al., 2023.[66]

Masculinist norms support indifference, even eco-violence, towards nature,[62] and concern for global warming is thus treated as a 'women's issue,' dismissed as less relevant than issues given higher priority within a masculinist logic, like economic growth.

Second, a masculinist ethos prescribes a general disregard for women as well as other groups deemed feminized, all of whom are relegated to lower status in patriarchal systems. Thus, white men espousing a masculinist ethos may readily dismiss the heightened vulnerability to the impacts of global warming among these groups. For example, in one study climate denial among conservative males was found to be associated with xenoscepticism: scepticism regarding issues faced by racialized minorities[63]. Third, because a masculinist logic shuns communitarianism and privileges individual autonomy, climate response options are vehemently opposed, as many of those options prioritize communities over individuals, and call for the redistribution of resources.

Fourth, at its core, a masculinist denial of the reality of climate change rests upon the objectification, exploitation, and ultimately the erasure of nature, alongside women, queer, black, brown, and Indigenous peoples, supported by a dualistic, hierarchical worldview. Fifth and finally, as members of a dominant group, men embracing masculinist identities will be particularly sensitive to perceived threats to their group and will escalate defensive reactions, while derogating out-groups.[64] When feeling threatened, as many privileged men are by progressive responses to climate change, reacting aggressively and even violently is deemed appropriate according to a (toxic) masculinist logic.[65]

Much as many of its adherents bluster on about their superior intellect, as with every other behaviour bucket described in this chapter, denial is more than anything an emotional response. As Katherine Hayhoe[67] put it, climate denial is not a cause, it's a symptom. It is a symptom of an unwillingness to acknowledge one's vulnerability, and culpability, and a deep-seated ontological insecurity that is layered upon other social and economic trends that have likewise served to rend into shreds a worldview that has comforted and justified stasis and privilege. Emotions like shame feature strongly, as does fear of ostracism from one's in-group, and fear of change generally, but also outrage about threats to status: in other words, autonomy.

Withdrawal

That behavioural response I refer to as withdrawal in some ways has a lot in common with denial. As with denial, withdrawal represents a heightened emotional reaction to information about climate change. Rather than reject the science, however, withdrawal refers to an intentional distancing from the subject of climate change among individuals who have difficulty coping with the high level of anxiety that can result, suggestive of some level of cognitive-emotional overload. Perhaps these individuals, were they to find themselves immersed in different social groupings that support climate denial, would become deniers themselves. Or, with sufficient emotional support, perhaps they would become activists. Without the availability of and susceptibility to denial narratives, or emotional support to favour action, withdrawal becomes the most likely course.

Alarm, even anxiety, is a perfectly reasonable reaction to news about the climate emergency, and among many can stimulate 'problem-focused' coping strategies, seeking to address the problem that is causing the anxiety. But when the threat posed is perceived as too complex and unwieldy to address, either due to the nature of the problem itself, one's personal capacities, or both, anxiety is more likely to favour 'emotion-coping' strategies, palliative responses, many of which manifest as withdrawal.[68] Many people turn away from anxiety triggers, including avoiding learning important information, as a way to steer clear of psychological discomfort.[69] For example, I have a heightened emotional response to child abuse, so much so that I often turn away from news, literature, or film in which child abuse is featured. This of course means I would not be the best candidate for a position in a social services agency that supports child abuse survivors, despite the fact that it is an issue I care about deeply.

In order to discuss withdrawal as a response to the climate emergency, we need to start with a general discussion about anxiety and the emerging scholarship on the phenomenon of climate anxiety. Anxiety is a mental state in which individuals direct heightened negative emotional attention towards anticipated future events or situations. Anxiety can affect one's behaviour, particularly when it causes cognitive-emotional impairment, perhaps in the form of difficulty sleeping or concentrating, and as functional impairment, such as a declining ability to achieve daily tasks, like preparing meals and getting to work on time.[70] Anxiety is often depicted in western societies as a malady. In other words, it falls under the category of unreasonable, irrational reactions, a disorder, one that warrants therapeutic intervention, so that one can return to routine, socially acceptable functioning. However, I think it is important to differentiate the mental state from the behavioural outcomes. Certainly cognitive-emotional impairment and functional impairment are horribly debilitating places to find oneself, with substantial impacts on the individual, and also their family, friends, and co-workers. But the mental state itself, directing heightened emotional attention to a threat, when that threat is the real deal, is not necessarily a bad thing. To the contrary, it is a crucial, inherited, survival-promoting trait. To 'manage' our climate alarm in the manner that we 'manage anxiety,' in other words to seek to eliminate the negative emotions, would amount to prevention of that very trigger that allows for an appropriate reaction to the serious climate threats we face.

The emergence of anxiety in response to the climate emergency is thus not only a reasonable emotional and cognitive reaction; it is an important route towards pro-climate action. Global warming is associated with high levels of uncertainty, and existential threats to the lives and lifestyles of many, and to ecosystems. As well, reckoning with our lack of progress despite full awareness of the scale of the problem is highly distressing not to mention enraging. And, as the term implies, it is global in scale. These are all certainly anxiety-inducing features of the climate emergency. Some degree of alarm is most likely necessary to motivate action, except for those few who simply perceive positive opportunities to themselves in engaging in pro-climate action and thus bypass the worry phase altogether (I know a few energetic, ever-positive people who would fit into this category). So, climate anxiety can be highly adaptive, a source of action motivation.[71] Anxiety can also be infectious, as with shared emotions generally. Charles Ogunbode and colleagues[72] showed that the level of reported climate anxiety is directly correlated with the number of people in one's social networks who also express distress about global warming.

The potential for negative behavioural outcomes, on the other hand, is another matter. For many, anxiety about climate change may be layered upon additional stressors, or be expressed by individuals who, for one reason or another, have limited emotional coping capacity. This situation favours emotion-coping responses,[73] seeking ways to ameliorate the negative feelings rather than confront the anxiety-inducing problem. Such a response may well be adaptive in the short term if you just do not have any more fuel in the tank and you need what's left to deal with a lost job, or harassment. But such responses are clearly maladaptive in the aggregate, and maybe also for the individual too, if it translates into lack of attention and hence a failure to adapt to the personal threats posed by the impacts of global warming.

Empirical studies of climate anxiety are small in number, but emerging rapidly. Among those studies, younger generations appear to be especially susceptible.[74] Several studies also show, not surprisingly, that individuals who have strong biospheric orientations—care deeply about the environment and value it for its own sake—are more likely to experience climate anxiety. Interestingly, exposure to climate change information does not in itself appear to be a consistent trigger. One recent study indicated that higher levels of reported climate knowledge were *negatively* related to climate anxiety,[75] while other studies show the exact opposite.[76] Ultimately, research suggests that only a minority of individuals who are worried about climate change show indications of clinical anxiety,[77] and the strongest predictor of climate anxiety is general anxiety[78]: individuals whose worry about climate change descends into anxiety are more disposed to anxiety in general.

All of these very recent findings are of interest, and important, but should be received with the understanding that this field of research is very new, and researchers are still working out the theoretical and methodological details. One particular weakness of this research record to date is that a distinction is not often made between a state of anxiety, and the different forms of behavioural response that result, leading to mixed results regarding the correlation between climate anxiety and pro-climate action. Some studies find a positive association between climate anxiety and pro-climate action, suggestive of an adaptive threat response,[79] while others instead find no association between climate anxiety and action, suggestive of a withdrawal response at least among some.[80] In one rare international comparative study, Ogunbode and colleagues[81] identify national variations in the degree of correlation between climate anxiety and climate action. While a positive association was observed in most countries in the study, including many European countries as well as Australia, Brazil, and Russia, no such association existed among participants living in China. Kim-Pong Tam and colleagues[82] found national differences in expressions of climate anxiety but not its bearing on action; in their study, Chinese and Indian participants indicated higher levels of anxiety than others in the sample, but anxiety was positively correlated with action in all countries sampled.

However, the very manifestation of global warming's impacts can tax our emotion-cognition capacities to such an extent that withdrawal can be the result. Rapid changes in ecosystems can lead to ecological grief and solastalgia—the deep sadness one feels when landscapes to which they are attached experience abrupt, substantial change, such as when the forested landscape out your front window burns to the ground, the lake your family has visited every summer dries up due to drought, or the medicinal herbs your family has used for generations no longer grow where they used to. Personal experiences with extreme events constitute traumas that are more difficult to cope with the weaker the social support systems one has access to. The increase in frequency and intensity of localized climate-attributed extreme events can directly generate anxiety, depression, and PTSD, particularly among marginalized groups with limited recovery resources, and groups particularly attached to their local ecosystems.[83]

The direct psychological impacts of fires, floods, and extreme heat are not the same thing as climate anxiety, but it can induce negative mental health outcomes that prime an individual towards a withdrawal response. Some people exposed to environmental risks tend to use emotion-focused coping strategies, rather than problem-focused ones. One particular study focused on coastal flooding found that residents who have difficulties regulating emotions, and those with a high level of place attachment, were most likely to use emotion-focused coping strategies.[84]

Stuck

Quickly becoming the largest behavioural group, those who are stuck—people who simply feel inert—are aware and concerned about global warming, and are not experiencing compromises to their mental health that support withdrawal, but nonetheless are just not taking action. In other words, they are stuck. Much research confirms that on the whole, the more one worries about climate change, the more likely one is to take action. But there are far too many reasons why we might not act in response to the concerns we have. Many individuals who value environmental wellbeing do not take personal actions that would reduce their ecological footprint, describing what has been called an environmental value-action gap. Nietzsche may have been right when he postulated that we are far better at collecting knowledge than we are at contemplating our role in the world.[85] There are several things that may be going on here, but efficacy is the big one. Do you believe you can make a difference? Do you believe that collectively *we* can make a difference? Numerous studies indicate that for many of us the answer to one or both of these questions is no, and lower perceived efficacy, not surprisingly, reduces engagement in pro-climate action.[86] *Why bother??*

One's personal sense of efficacy in the face of the climate emergency can be easily squashed. Just running some numbers of your own household emissions in comparison to aggregate emissions is sobering enough. Even if I were to achieve the herculean task of bringing my household emissions down to zero, my family's efforts would have an infinitesimally small effect on the climate. But many also just do not have the ability to adopt the changes suggested, due to where they live, or their income, leading to a permanent state of cognitive dissonance, a terrible mental space to find oneself. If your access to public transit is limited, and you cannot afford an electric vehicle, then you take the gas guzzler out of the garage to get to work and do your errands, feeling guilty, grumpy, and powerless to make the changes that would be more in accordance with your values.

Regardless of one's sense of personal efficacy, if they do not have confidence that collectively we can generate desired change, perhaps because they presume that institutions like governments will fail to perform as expected, this may lead to a fatalistic course of inaction as well.[87] I may understand perfectly that food waste is a significant contributor to greenhouse gas emissions, but what is the point of attempting to utilize every scrap of food in my kitchen when the dumpsters outside grocery stores are filled to the brim with perfectly good food every night?

Believing that others are not as concerned as you—a perception that you are out of step with your in-group—is another factor facilitating an inert response. This belief can feel deflating, and stifle motivation to act. Many in the U.S. underestimate the levels of acceptance and concern about climate change among others. Gregg Sparkman and colleagues,[88] for example, found that while 66–80% of surveyed Americans support major climate change mitigation policies, those respondents estimated climate policy support among others to be just 37–43%. Americans also generally over-estimate the degree of partisanship over the issue.[89] If you think you're alone in your concern, or that political parties will never see eye to eye and therefore our elected officials can't be counted on to do the right thing, then you might just feel quite discouraged about even discussing the climate emergency, much less taking action to address it.

This may be a type of self-fulfilling prophecy. In order for climate change information or even the direct experience of climate change impacts to motivate action, individuals need to have an opportunity to discuss their experiences with others, feel validated, and be a member of a group. But many of us belong to social groups in which we don't often talk about climate change, even after climate-attributed extreme events. In a rare study observing community discussions after an extreme event experience, Boudet and colleagues[90] found that such events led to discussions of climate change in only a small minority of affected communities, with most attention instead directed to emergency response management and economic recovery.

Conclusions

Each of these inaction behaviour responses reflects emotional pathways that in one way or another inhibit an adaptive response to the climate emergency. I am inclined to believe that inaction can be attributed in very many cases to defence of social belonging, and/or a compromised sense of self-determination, enabled by efficacy and autonomy. As discussed in Chapter 4, we each have certain innate, basic psychological needs, including relatedness, or belonging, a sense of competence, or efficacy, and a sense of personal autonomy, or the ability to choose a course of action without being compelled to do so. When these needs are met, a positive condition of mental wellbeing results, and this in turn allows for the capacity to cope with stressors and threats in a constructive way.

On the other hand, people do change! Apathy, denial, withdrawal, and being stuck are all transient behavioural states, particularly because they are emotionally unstable. Ballew et al. (2022) even found that many individuals can change their mind about climate change, from denial to acceptance, with increased exposure to information from trusted sources. Survey research conducted in Australia[91] showed that climate scepticism declines, ever so slightly, as the previous year's global average temperatures increase. So yes, there are many pathways to inaction, perhaps more than to action, but that too may be beginning to change. On to the next chapter.

Notes

1 Olson, *The Logic of Collective Action*.
2 Brick, Bosshard, and Whitmarsh, "Motivation and Climate Change."
3 Lidskog et al., "Cold Science Meets Hot Weather."
4 Fisher, *Human Communication as Narration*.
5 Bransford, National Research Council (U.S.), and National Research Council (U.S.), *How People Learn*.
6 Morris et al., "Stories vs. Facts." P. 21.
7 Albright and Crow, "Beliefs about Climate Change in the Aftermath of Extreme Flooding"; Denny, Marchese, and Fischer, "Severe Weather Experience and Climate Change Belief among Small Woodland Owners"; Howe et al., "How Will Climate Change Shape Climate Opinion?"
8 Davidson, Fisher, and Blue, "Missed Opportunities."
9 Ogunbode, Doran, and Böhm, "Individual and Local Flooding Experiences Are Differentially Associated with Subjective Attribution and Climate Change Concern"; Osberghaus and Fugger, "Natural Disasters and Climate Change Beliefs."
10 Zanocco et al., "Personal Harm and Support for Climate Change Mitigation Policies."
11 Sugerman, Li, and Johnson, "Local Warming Is Real."
12 Lynch, "We Have Never Been Anti-Science."
13 Skurka et al., "Pathways of Influence in Emotional Appeals: Benefits and Tradeoffs of Using Fear or Humor to Promote Climate Change-Related Intentions and Risk Perceptions"; Skurka, Myrick, and Yang, "Fanning the Flames or Burning Out?"; Diamond and Urbanski, "The Impact of Message Valence on Climate Change Attitudes"; Morris et al., "Stories vs. Facts."
14 Bloodhart, Swim, and Dicicco, "Be Worried, Be VERY Worried."
15 Passyn and Sujan, "Self-Accountability Emotions and Fear Appeals: Motivating Behavior"; Witte and Allen, "A Meta Analysis of Fear Appeals: Implications for Effective Public Health Campaigns."
16 Kemkes and Akerman, "Contending with the Nature of Climate Change."
17 Chapman et al., "Climate Visuals."
18 Hassol, "The Right Words Are Crucial to Solving Climate Change – Scientific American"; Lidskog et al., "Cold Science Meets Hot Weather"; Moser, "Communicating Climate Change: History, Challenges, Process and Future Directions."
19 Wolf and Moser, "Individual Understandings, Perceptions, and Engagement with Climate Change."
20 Moser and Dilling, "Communicating Climate Change: Closing the Science-Action Gap." P. 168.

21 Luo and Yu, "Follow the Heart or the Head?"
22 Van Boven et al., "Feeling Close."
23 Conte, Brosch, and Hahnel, "Initial Evidence for a Systematic Link between Core Values and Emotional Experiences in Environmental Situations."
24 Giddens, "The Politics of Climate Change."
25 Tang and Chooi, "From Concern to Action"; Harth, "Affect, (Group-Based) Emotions, and Climate Change Action."
26 Tang and Chooi; Spence, Poortinga, and Pidgeon, "The Psychological Distance of Climate Change."
27 Wang, Geng, and Rodríguez-Casallas, "How and When Higher Climate Change Risk Perception Promotes Less Climate Change Inaction."
28 Shaw, Scully, and Hart, "The Paradox of Social Resilience"; Ogunbode et al., "The Resilience Paradox."
29 Shao and Yu, "Media Coverage of Climate Change, Eco-Anxiety and pro-Environmental Behavior."
30 Markowitz and Shariff, "Climate Change and Moral Judgement." P. 243.
31 Jasanoff, "A New Climate for Society." P. 248.
32 Wolf and Moser, "Individual Understandings, Perceptions, and Engagement with Climate Change."
33 Bosca, "Comfort in Chaos."
34 Bloodhart, Swim, and Dicicco, "Be Worried, Be VERY Worried."
35 Markowitz et al., "Compassion Fade and the Challenge of Environmental Conservation."
36 Haltinner, Ladino, and Sarathchandra, "Feeling Skeptical"; Haltinner and Sarathchandra, "Climate Change Skepticism as a Psychological Coping Strategy."
37 Brulle and Norgaard, "Avoiding Cultural Trauma."
38 Norgaard, *Living in Denial*.
39 Zaremba et al., "A Wise Person Plants a Tree a Day before the End of the World."
40 Aguilar-Luzón, Carmona, and Loureiro, "Future Actions towards Climate Change."
41 McCright and Dunlap, "Bringing Ideology In."
42 Benegal, Azevedo, and Holman, "Race, Ethnicity, and Support for Climate Policy."
43 Albright and Crow, "Beliefs about Climate Change in the Aftermath of Extreme Flooding."
44 Hamilton et al., "Flood Realities, Perceptions and the Depth of Divisions on Climate."
45 Weber, "Seeing Is Believing"; Uenal et al., "Climate Change Threats Increase Modern Racism as a Function of Social Dominance Orientation and Ingroup Identification."
46 Kemeny, Gruenewald, and Dickerson, "Shame as the Emotional Response to Threat to the Social Self: Implications for Behavior, Physiology, and Health"; Yeager, "Collective Shame in Climate Denial."
47 Yeager, "Collective Shame in Climate Denial." P. 2.
48 Arce-García, Díaz-Campo, and Cambronero-Saiz, "Online Hate Speech and Emotions on Twitter"; Colleoni, Rozza, and Arvidsson, "Echo Chamber or Public Sphere?"
49 Yeager, "Collective Shame in Climate Denial"; Veldman, *The Gospel of Climate Skepticism*.
50 Veldman et al., "Who Are American Evangelical Protestants and Why Do They Matter for US Climate Policy?"
51 Ferguson and Branscombe, "Collective Guilt Mediates the Effect of Beliefs about Global Warming on Willingness to Engage in Mitigation Behavior."

52 Mayer and Smith, "Multidimensional Partisanship Shapes Climate Policy Support and Behaviours."
53 Spence and Ogunbode, "Angry Politics Fails the Climate."
54 Merkley and Stecula, "Party Cues in the News"; Rinscheid, Pianta, and Weber, "What Shapes Public Support for Climate Change Mitigation Policies?"
55 Hahnel, Mumenthaler, and Brosch, "Emotional Foundations of the Public Climate Change Divide."
56 Jylhä and Akrami, "Social Dominance Orientation and Climate Change Denial"; Stanley and Wilson, "Meta-Analysing the Association between Social Dominance Orientation, Authoritarianism, and Attitudes on the Environment and Climate Change"; Uenal et al., "Climate Change Threats Increase Modern Racism as a Function of Social Dominance Orientation and Ingroup Identification."
57 Dhont et al., "Social Dominance Orientation Connects Prejudicial Human–Human and Human–Animal Relations"; Milfont et al., "Environmental Consequences of the Desire to Dominate and Be Superior."
58 Jylhä and Akrami, "Social Dominance Orientation and Climate Change Denial"; Duckitt and Sibley, "Personality, Ideology, Prejudice, and Politics"; Pratto et al., "Social Dominance Orientation."
59 Flynn, Slovic, and Mertz, "Gender, Race, and Perception of Environmental Health Risks." P. 1107.
60 Hultman and Pulé, *Ecological Masculinities*.
61 Daggett, *The Birth of Energy*.
62 Blenkinsop, Piersol, and Sitka-Sage, "Boys Being Boys"; Breunig and Russell, "Long-Term Impacts of Two Secondary School Environmental Studies Programs on Environmental Behaviour."
63 Krange, Kaltenborn, and Hultman, "Cool Dudes in Norway."
64 Uenal et al., "Climate Change Threats Increase Modern Racism as a Function of Social Dominance Orientation and Ingroup Identification."
65 Pulé and Hultman, *Men, Masculinities, and Earth*.
66 Davidson et al., "The Symptoms of Climate Denial: Tracing the Intersections of Denial, Right-Wing Extremism and Masculinity."
67 Bjork-James and Barla, "A Climate of Misogyny."
68 Stollberg and Jonas, "Existential Threat as a Challenge for Individual and Collective Engagement."
69 Shepherd and Kay, "On the Perpetuation of Ignorance"; Webb, Chang, and Benn, "'The Ostrich Problem.'"
70 Helm et al., "Coping Profiles in the Context of Global Environmental Threats"; Cruz and High, "Psychometric Properties of the Climate Change Anxiety Scale."
71 Hepp et al., "Introduction and Behavioral Validation of the Climate Change Distress and Impairment Scale."
72 Ogunbode et al., "Climate Anxiety, Wellbeing and pro-Environmental Action."
73 Barlow, *Anxiety and Its Disorders*; Clayton and Karazsia, "Development and Validation of a Measure of Climate Change Anxiety."
74 Galway and Field, "Climate Emotions and Anxiety among Young People in Canada"; Clayton and Karazsia, "Development and Validation of a Measure of Climate Change Anxiety."
75 Zacher and Rudolph, "Environmental Knowledge Is Inversely Associated with Climate Change Anxiety."
76 Bright and Eames, "From Apathy through Anxiety to Action"; Hickman, "We Need to (Find a Way to) Talk about … Eco-Anxiety"; Verlie et al., "Educators' Experiences and Strategies for Responding to Ecological Distress."
77 Whitmarsh et al., "Climate Anxiety"; Wullenkord et al., "Anxiety and Climate Change."

78 Asgarizadeh, Gifford, and Colborne, "Predicting Climate Change Anxiety"; Wullenkord et al., "Anxiety and Climate Change."
79 Sangervo, Jylhä, and Pihkala, "Climate Anxiety"; Whitmarsh et al., "Climate Anxiety"; Van Valkengoed, Steg, and Perlaviciute, "The Psychological Distance of Climate Change Is Overestimated."
80 Coffey et al., "Understanding Eco-Anxiety"; Clayton and Karazsia, "Development and Validation of a Measure of Climate Change Anxiety"; Kapeller and Jäger, "Threat and Anxiety in the Climate Debate—An Agent-Based Model to Investigate Climate Scepticism and Pro-Environmental Behaviour"; Wullenkord et al., "Anxiety and Climate Change."
81 Ogunbode et al., "Climate Anxiety, Wellbeing and pro-Environmental Action."
82 Tam, Chan, and Clayton, "Climate Change Anxiety in China, India, Japan, and the United States."
83 Findlater et al., "Integration Anxiety."
84 Navarro et al., "Coping Strategies Regarding Coastal Flooding Risk in a Context of Climate Change in a French Caribbean Island."
85 Anfinson, "How to Tell the Truth about Climate Change."
86 Bieniek-Tobasco et al., "Communicating Climate Change through Documentary Film."
87 Toivonen, "Themes of Climate Change Agency."
88 Sparkman, Geiger, and Weber, "Americans Experience a False Social Reality by Underestimating Popular Climate Policy Support by Nearly Half."
89 Ballew et al., "Beliefs about Others' Global Warming Beliefs."
90 Boudet et al., "Event Attribution and Partisanship Shape Local Discussion of Climate Change after Extreme Weather."
91 Hornsey, Chapman, and Humphrey, "Climate Skepticism Decreases When the Planet Gets Hotter and Conservative Support Wanes."

7 Pathways to Action, or Doing the Hard Thing

I love hiking. The longer the trail, the larger the elevation gain, the better. I love the views for sure, especially the ones to be had at sunrise in the eastern Sierras in California, my all-time favourite place to be with my pack and poles. But views aplenty are available from the comfort of my arm chair, in books, movies, and of course on the internet. Hiking offers so much more than a great view though. There is something that is just so exhilarating, so satisfying, about being *in* the view, feeling, hearing, and smelling it rather than just seeing it; and moving within it, following a challenging trail to its completion, with that refreshing, squeezed-out sponge feeling when you're done. Yet, for many people, such a course of action would sound preposterous. 'Why ever would you do that?' they are likely to ask. And it's a fair question—why expend some 2,000 calories, push my body to its physical limits, and risk injury or perhaps even death, to walk someplace I don't have to go,

Figure 7.1 I was sore for days after I took this hike with friends to the top of Sarrail Ridge in Kananaskis Country, Alberta. Was it worth it? Absolutely. Photo taken by author, whose dusty boot is sticking up, bottom centre. Friends featured graciously permitted its reproduction here.

only to turn around and walk back to where I started, with nothing to show for it, other than some amateur photographs and sore muscles? Why indeed do we ever do the hard thing? For love, of course.

The premise of this book is that higher levels of engagement in pro-climate action are necessary to effective mitigative and adaptive responses to the climate emergency, and upscaling engagement requires attending to emotionality: overcoming the emotional barriers to adopting and sustaining new practices including, importantly, collective action. Willingness to invest one's limited energy, time, and money, potentially facing social backlash, departing from established and routinized lifestyles and potentially belief systems as well, are unlikely to be pursued on the basis of utilitarian expectations alone. Mancur Olson, in *Logic of Collective Action*,[1] got that much right: the costs of collective action in pursuit of public goods are inevitably higher than the rewards—from a utilitarian perspective. What Olson got wrong was the presumption that self-interest is the only plausible motivation for collective action. People also choose courses of collective action as an expression of care for things being threatened, including care for others, care for ideals (like justice), or simply for the non-material personal rewards, in the form of belonging, offered by collective engagement itself. Such investments are emotional, not instrumental. And yet, these decisions are also not taken lightly: to reject the prevailing institutional ethos that prescribes stasis, routines, and non-reflexivity is enormously costly—it may well represent a proverbial mountain much higher than Sarrail Ridge, featured in the photo above—even if only done privately, but particularly if done so publicly. As such, investing in the institutional changes necessary to low-carbon transitions requires the opportunity to experience sufficient positive emotional rewards to counter the costs.

Upscaling pro-climate action thus demands attention to the importance of emotions in disrupting inaction pathways to support a shift towards pro-climate action, and then sustaining those commitments. While pro-climate action includes a wide array of personal and collective projects, all forms of action will be associated with emotion-cognition pathways that depart significantly from inaction pathways. In this chapter, we explore the unique attributes associated with pro-climate action pathways. I will first revisit a discussion that was introduced earlier, of agency and reflexivity, and the bearing of socio-cultural, sociopolitical, and economic structures on agency and reflexivity, which manifests as intersectionality. Next, I'll clarify what we mean by 'pro-climate action,' including a variety of projects all of which share in common a deliberate effort on the part of an individual to take action intended to contribute to the mitigation or adaptation of climate change. I then explore alarm triggers, and how they land differently for people who choose action over inaction. However, while alarm triggers are a crucial first step towards pro-climate action, it bears repeating that action does not inevitably, or even typically, follow on climate concern. Therefore, after discussing alarm triggers, we need to attend to what I call the 'mechanisms of agency,' the emotional and cognitive conditions and processes through which an individual chooses to act on the basis of their concerns: the intervening factors that enable the translation of alarm into action.

A Brief Recap on Reflexivity and Power

We discussed reflexivity at length in Chapter 3. Of particular interest to this chapter is a particular type of reflexivity that Margaret Archer termed 'meta-reflexivity,' which describes an inclination to contemplate and problematize our circumstances, according to a strong personal value compass, and then pursue courses of action that have the intention of confronting those problematized circumstances. In other words, meta-reflexives have a particularly central role to play in the pursuit of intentional, transformational social change, because they are the ones willing to challenge current institutional and structural orders. Personal engagement in projects intended to address global warming, and particularly collective action projects, demands meta-reflexivity. The alarm triggers and mechanisms of agency discussed below can be conceived as different components of our reflexivity towards the climate emergency.

There is another element to reflexivity, or more precisely, another interpretation of meta-reflexivity, that I consider to be particularly important to pro-climate action, and to our collective responses to the climate emergency generally: our predispositions towards change. The sign in the photo below, posted in early October on the grounds of the school near my home, is a prompt to embrace change. But why do we need to be prompted? Why do we need to be reminded that change can be a good thing? Of course, change is not always and inevitably beautiful, as the sign suggests—there is nothing beautiful about losing a leg due to a land mine—but there are certainly some redeeming features about change. Sometimes change is just plain necessary in our personal lives—ending an unhealthy romance, say, or recovering from an addiction. And sometimes, as in our current moment in the midst of a climate emergency, collective, transformative institutional change is definitely, absolutely, necessary. Yet, I have often been struck by the fact that, while in evolutionary terms we would appear to be among the most adaptive species on the planet, many of us really and truly do not like change! We are creatures of routines and habits; habits which are readily formed and yet only broken with difficulty. In many ways, this is what meta-reflexivity is all about: A recognition of the need for change, a willingness to embrace change, and to go a step further and deliberately pursue change.

Photo taken by author, Edmonton Alberta, September, 2023.

We also cannot contemplate processes of reflexivity without accounting for power. Agency is literally the power to act with intention according to one's personal vision. Agency is thus intricately related to efficacy, which we will explore further below, and they both are guided by social relations of power. There are many sources of personal disempowerment that may surface regardless of one's positionality. However, that positionality has a very large effect: the effects of gender, the colour of one's skin, sexual orientation, Indigeneity, these are all social categories that have been ordered in our modern social structures hierarchically, such that certain categories—white, male, settler, English-speaking, etc.—are granted higher status by default. Among the most obvious of dimensions of intersectionality that combines with these other social categories is class—in the form of access to material resources, and the social privilege and political influence that wealth provides. Just as relevant are the non-material dimensions of class position, such as the empowered agency cultivated among the upper classes; that sense of expectation of responsiveness from others, of reward for one's efforts. Personal anecdote: I do not come from a family of privilege; my dad was a schoolteacher, and once my parents divorced, my mom entered the workforce as an entry-level administrative assistant. On my financially precarious journey through university, I never applied for anything other than student loans and financial aid; receiving a scholarship was just too far outside my perceived realm of possibilities. On the other hand, I began to make friends from families of privilege, friends who would apply for everything, and be outraged when they weren't successful. This sense of entitlement, of expectation of positive outcome, these are prefaces to taking action, to agency. On the other hand are the large differentials in the likelihood for personal life experiences that can compromise one's agency afforded by one's positionality. Reflexivity is enabled by mindfulness, by having the mental energy for deliberation. Traumatic experiences, and everyday stress, as discussed in previous chapters, serve to sap that mental energy. All in all, those in oppressed and marginalized positions have a much higher hill of emotional baggage to climb towards engagement in personal and collective projects to support socio-ecological change. This is a travesty, because these are precisely the people whose perspectives and contributions we need the most.

Pro-climate Action Types

There are four basic categories of climate action, capturing a wide variety of projects, including personal lifestyle projects; personal politics; collective lifestyle projects; and finally collective politics (Table 7.1). Although one might engage in activities that have positive benefits in the form of climate mitigation or adaptation for other reasons—even climate-denying farmers, for example, may engage in soil enhancement or wetland protection, which have climate mitigative benefits[2]—in this chapter we are particularly interested in actions undertaken with the deliberate intention to respond to climate concern. It

134 *Pathways to Action, or Doing the Hard Thing*

Table 7.1 Basic categories of climate action

Personal lifestyle projects	Collective lifestyle projects
Definition: Commitments to change personal and household-level practices understood to contribute to global warming, particularly but not exclusively related to material consumption. **Examples**: Adopting low-carbon technologies like solar panels and heat pumps; changing consumption practices, like choosing vegetarianism or foregoing long-distance flights; pursuit of careers in climate-related fields; choosing to have fewer children.	**Definition**: Pursuit of group-level projects designed to support lifestyle changes, most often but not exclusively undertaken at the local level. **Examples**: Starting or joining food/car sharing initiatives; organizing community energy co-ops; transition Towns and experiments in communal living; contributing to information-sharing virtual networks such as prosumer tips.
Personal politics **Definition**: Political acts that are undertaken individually. **Examples**: Prioritizing climate policy support in selection of candidates; petition signing; letters to elected officials; pro-climate social media engagement.	**Collective politics** **Definition**: Engagement in collective action projects that have the specific goal of pursuing change through politics. **Examples**: Membership in environmental organizations; direct participation in climate-related social movement activity such as protests; running for office.

is the intention, in other words, or the *means,* rather than the outcomes, or ends, that are of interest here. My purpose in this chapter is to focus on how emotions feature in personal commitments to act, rather than assert any evaluative judgement on the efficacy of the different actions that may be chosen. All four of these categories have positive outcomes, although each on its own is most likely necessary but insufficient to support transformational social change. One necessary ingredient to pro-climate action pertains to all four, and may go a long way to influence the selection of pro-climate action projects across the four categories: opportunity. Is there a pro-climate candidate in your riding? Are there bus routes available? Are there other members of your social networks with similar levels of climate concern to work with?

The Trigger

By what means does a concern motivate the planning and execution of action, and how do emotions feature in such deliberations? Once again, we begin at the beginning—that point at which personal alarm is triggered. First, information about the threats posed by global warming can trigger alarm bells when one feels personally threatened. We unfortunately face a relative dearth of research on this most direct of triggers, because research indicates that individuals tend not to perceive global warming as a personal threat, a finding that I believe cannot be separated from the fact that the vast majority of this empirical research has been conducted among middle class residents of advanced industrial nations. Alarm can also be triggered for many in response to care for threatened others with whom we identify, even if we do not perceive a personal threat. But first, to elaborate upon what is perhaps obvious, these triggers do not feel good!

Although we all prefer to experience the pleasure of positive emotions, negative emotions such as fear are far more likely to get us out of our proverbial chairs to do something about a problem.[3] Negative emotions, after all, are an integral part of our emotional palette because of their ability to get our attention in order to respond to potential danger, and they do indeed tend to command our attention in ways that positive emotions do not.[4] It should thus come as no surprise that empirical studies of climate concern repeatedly show that negative emotions like worry, fear, guilt, and anxiety are strongly correlated with concern about the climate emergency, to a much greater extent than positive emotions.[5] These negative emotions, however, are necessary but not sufficient to support climate action.[6] One crucial intervening factor precipitating climate concern, as discussed in the last chapter, includes the emotional capacity to sit with those negative emotions, rather than pursue denial, or withdrawal.

The climate emergency as a subject of attention, however, can only trigger alarm when connections are made between global warming and the things that matter to the perceiver: one's 'objects of care.'[7] The gravity of this seemingly simple observation made by Wang and colleagues is worth some arm waving. As obvious as the need for this connective tissue may seem, much of the content of climate communications remains focused on the science—on increments of warming, on carbon molecules, and shrinking ice sheets, all reported in equally obscure terms of p-values and error margins—which, for those lacking in climate science training, in other words the vast majority of people on the planet, would appear to have little relevance to daily living.

Those connections between global warming and the things that matter, however, are beginning to surface. The first and most obvious form is as a personal threat, mentioned above. There is no question that feeling personally threatened is a strong motivator for actions intended to re-establish security, and the impacts of global warming have already begun to pose such acute personal threats to instigate proverbial 'fight or flight' responses taking a variety of forms, like migration, or engaging in conflicts over water supplies.

Youth climate activism is also driven at least in part by concerns among many young people today for their own futures. Personal financial interest, particularly among those who cannot afford NOT to take an interest in their finances, is also an immediate driver of behaviour change, when such changes are feasible. Many households adjust their consumption by necessity in response to an increase in the price of gasoline or beef, in household heating and electricity, and while several factors affect the prices of these household necessities, direct global warming impacts on supplies, as well as climate policies, can play a significant role. On the flip side, a reduction in the price of emerging green technologies like heat pumps and electric vehicles can be expected to increase their uptake, which is why the governments of some countries, like Canada and the U.S., have begun to offer substantial rebates on these purchases. Such forms of economically induced behaviour change can certainly contribute to climate mitigation (although I am by no means suggesting price increases for the poor as an effective climate policy), but prices and rebate offers can be volatile, unreliable drivers of household consumption—note my earlier critiques of rational-utilitarian models of human behaviour. In any case, neither economically driven changes in consumption behaviour, nor actions in response to personal threats (with the exception of youth climate activism), constitute actions motivated by a desire to avert a climate emergency, which is the main interest of this chapter.

Research suggests, to the contrary, that material threats and opportunities are not, on the whole, the main driver of concern about the climate emergency. Perhaps they should be more so than they are, considering the estimated economic costs of runaway climate change are astronomical, far outpacing the costs of societal transition towards carbon free futures. So, if not money, what are the values, the cares, the imaginaries that are linked to climate concern? Perhaps not surprisingly, a substantial research record indicates that climate concern is linked not to self-interest, but to care for others, in particular others outside of our immediate in-group, referred to as altruistic values. This includes care for, and a feeling of connectedness to, nonhuman others, and Nature herself, referred to as biospheric values.[8] The important work done by these value sets is in their capacity to override psychological distance. In other words, while increasing proximity of threat—perceiving threats to one's family, local community, and neighbourhood—tends to attract attention and by extension concern, the reverse is also true: feeling strong emotional attachment to objects of care, even when those objects are not proximal, can override the inattention that psychological distance supports.[9]

If only we all embraced altruistic and biospheric values! While we all most certainly have the capacity for altruism, and we are all indelibly connected to nature and also have the capacity to recognize, and care about, those connections, cultural and social structures impose vastly different levels of reward and sanction for cultivating those values. As outlined by Conte and colleagues,[10] the centrality in Western cultures of self-enhancement values (concern for wealth, social power, personal pleasure) versus self-transcendent values (including altruistic and biospheric values) goes a long way toward explaining complacency towards environmental and climate threats.

Despite these caveats, there is no question that the number of people across the world who worry about climate change is on the rise, and this is particularly the case for youth.[11] This alarm is also a strong predictor of pro-climate action,[12] so its escalation is certainly noteworthy. However, as we will explore in detail in the next section, alarm is by no means sufficient on its own to motivate action. What is needed is the presence of one or more mechanisms of agency: processes that enable and empower an individual to act on their alarm. To fight, in other words, instead of fleeing or freezing.

Mechanisms of Agency: From Alarm to Action

There is a lot going on here, which I have divided into seven distinct, but inevitably inter-related, mechanisms of agency; not all are necessarily emotions in and of themselves, but are nonetheless closely inter-related with our emotionality. These mechanisms are essential ingredients to support the transference of concern to action, many of which resonate with a number of the principles of emotionality described in Chapter 4. These include: norms of responsibility; efficacy; belonging; empathy; complexity thinking; future-gazing; and hope.

Norms of Responsibility

Strongly held norms that prescribe responsibility represent a particularly important mechanism of agency, referring not—or at least not solely—to an altruistic sense of responsibility to act in response to consequences on others, but more precisely a sense of *accountability* for one's own actions, or even the actions of one's in-group. Some empirical research shows quite clearly that climate concern is most likely to motivate pro-climate action when in the presence of personal feelings of responsibility,[13] a responsibility that can be extended towards distant others.[14] Not all socio-cultural settings, however, prescribe norms of responsibility. Even in the presence of strong self-transcendent values, in other words, cultural norms may prescribe respect for and deference to authority, or the minimization of attention to one's personal impact in favour of a focus on the collective.

Norms of responsibility can, among other things, generate a sense of guilt that surfaces with an awareness that one's own behaviours played some role in causing negative impacts for others. Despite the fact that guilt is a negative emotion that we would all rather avoid, feelings of guilt in response to a sense of responsibility for contributing to global warming has been shown in repeated studies to be a strong predictor of climate policy support, and engagement in different forms of pro-climate action, performing consistently better than positive emotions like pride.[15] That attribution of responsibility, moreover, can even be extended to one's in-group as a whole, such that one can feel guilty by association with a specific group identified to have had a disproportionate role to play in the production of greenhouse gas emissions.[16]

There has been some particularly interesting new experimental research indicating that invoking norms of responsibility that are specific to different groups can favour pro-climate action, even among groups not inclined to do so. For example, as discussed in Chapter 4, the role of norms and values are equally relevant to the feelings and actions of both liberals and conservatives, but the norms and values embraced by these groups are quite different. And while climate advocacy statements most often speak to liberal norms and values, for equity for example, pro-climate action can potentially be framed in ways that align with conservative values as well. Wolsko, Ariceaga, and Seiden[17] tested this idea, and found that conservatives can be persuaded to engage in pro-climate action, when doing so is framed in a manner that appeals to the ideals of tradition and patriotism. More recently, Clark and Adams[18] found similar results: conservative research participants in their study exhibited a notable shift towards favouring pro-climate action, when such actions were framed as a matter of obeying authority, and patriotism.

Efficacy

Efficacy refers to one of those basic human needs for emotional wellbeing discussed in Chapter 4. The concept of efficacy is similar to, and often confused with, the concept of agency itself, but there is one important difference. Agency is an outcome of the actions of an actor, an outcome that is at least ostensibly observable. Efficacy on the other hand is a feeling, conscious perception on the part of the actor. Efficacy reflects, first of all, an awareness of one's *self*-determination—the expectation that I can try out for the soccer team if I want to, regardless of my income, gender, or race. This form of efficacy is what Ann Bostrom and colleagues[19] refer to as personal self-efficacy, which they differentiate from another form of efficacy, response efficacy. Response efficacy is the perceived ability to make a difference, to have effect—it is the confidence that, if I complain to the boss about the broken coffee machine, it will get fixed; if I bring my concerns to the City Council meeting, I will be heard. This desire for a sense of purposefulness, this assurance that I exist by virtue of the fact that the world around me responds to my actions, is easy enough to see even in toddlers, and frustrations of efficacy generate acute emotional reactions, anger mainly, to perceived unfairness to oneself. By extension, when combined with empathy, which gives us the capacity to feel what others feel, we are inclined to feel enraged when we observe injustices imposed upon others as well. Efficacy is thus closely linked to our innate sensitivity to fairness. In addition to these forms of individual efficacy, collective efficacy is particularly important to climate action. Collective efficacy refers to confidence in the collective's capacity to change, whether that collective refers to your local community, or the United Nations.[20] Not surprisingly, perceptions of efficacy frequently emerge as strong drivers of different forms of personal and collective pro-climate action in empirical research.[21]

This close association to fairness, or heightened responsiveness to perceived injustices and the anger and rage that often ensue, ironically allows for action, collective action in particular, even—especially—among groups that are notably lacking in political power. In other words, among members of groups that one might presume would express very low levels of efficacy. Work by Indigenous, Black, and feminist scholars has been particularly valuable to our understanding of collective action engagement by those who have no reason to expect success, whose agency is obstructed by structural forces, and for whom in many cases engagement in collective action is downright dangerous. Some feminist and Indigenous scholars[22] point in particular to the power of anger and rage in response to conditions of obstructed agency, marked by decades, centuries even, of exploitation, oppression, repeated lies, and violence perpetrated upon women, the poor, and Black, brown, and Indigenous peoples. Marisol LeBrón's[23] powerful analysis of women-led protests in Puerto Rico after the devastation of Hurricane Maria offers a valuable window into this interplay between efficacy and fairness. Rage at their obstructed agency, for these Puerto Rican activists, *became* their source of efficacy; the fuel that energized the movement. Embodied symbolically in the movement's master narrative centred on *coraje*—a Spanish word meaning rage as well as courage—that rage became not just a source of empowerment in their confrontations with colonialism and state violence, but also an opening for building alliances, and envisioning transformative change.

Belonging

As the pandemic made painfully clear to so many of us, our social connections, our confidence that we belong to others and they to us, is everything. In many ways, this is what gets us out of bed in the morning, and it is ever-present as we identify problems, contemplate responses to those problems, and ultimately enact personal projects. In other words, belonging, along with so many other aspects that invoke emotions, is absolutely in play in reflexivity. Like efficacy, belonging is a basic human need, one that transpires in the form of many emotions, but three in particular: shame, pride, and empathy.

Among those individuals who come to find themselves questioning— questioning the beliefs, norms and practices of their in-groups—the first temptation is to keep quiet, to avoid the shame that may result from group confrontation. But those individuals may also begin to peek beyond group borders, to see if there might be a different group that might be more in alignment with one's shifting personal identity. Pursuing membership in new groups and social networks can then lead to exposure to new information, new norms, and values that validate rather than shame. Re-orientation of one's social networks thus creates openings for imagining new beliefs and practices, and those networks simultaneously provide the social support—a movement from shame to pride—for embracing those beliefs and engaging in those practices. My research team and I discovered this first-hand, within a group that is known for upholding particularly rigid group boundaries: rural farmers.[24] In that study, we had the privilege of interviewing many Albertan farmers who made the shift from conventional farming practices (monocropping with heavy use of chemical inputs) to a variety of ecologically beneficial production practices. These farmers discussed the ostracism they felt among their neighbours when they voiced their critiques of conventional farming or displayed new practices in their fields, but their sense of belonging and enthusiasm for their new identity as ecologically beneficial farmers changed completely once they became immersed in the growing international social network of other farmers also trying to support more sustainable, and climate-neutral, forms of agriculture.

The absence of belonging—or alienation—is nothing short of debilitating, and is a state of being that one may very well find themselves when pursuing pro-climate action, or any other form of personal or collective project for that matter. Laura Jenkins[25] associates alienation with 'a feeling of isolation, dejection, loneliness, meaninglessness, despair, loss of autonomy and powerlessness and can be expressed differently—hostility, lethargy, shame and so on.' What is more, Jenkins continues, certain people are far more likely to experience alienation than others: 'alienation is often, but not always, connected to oppression and poverty in terms of income, capabilities and wellbeing. ... Thus, the poorest, most marginalised and alienated in society are among the least likely to engage politically.' But positionality is not the only, or determinative factor at play. Anyone, even members of privileged positionalities, who seek a course of beliefs and actions that conflict with the dominant norms and beliefs of their in-group, can feel alienated. Those under such conditions with the means to do so may feel highly motivated to seek out new places of belonging. This is a situation faced by many climate activists, particularly youth, as I found in my work with colleague Carrie Karsgaard.[26] Youth who find that their deep concern about the climate emergency is out of step with their families, friends, and teachers are likely to avoid expressing their concern, which only deepens their sense of isolation. On the other hand, once such isolated youths find each other, a new community is born. Participation in the climate movement itself can offer rewards in the form of belonging.

Let's move into a discussion of the specific emotions in play within groups, beginning with a brief look at how shame and pride have played out in recent research on attitudes and behavioural responses to the climate emergency. Although some research discusses shame's association with denial,[27] there has not yet been a wealth of empirical research focused on shame as an inducement to pro-climate action, with the exception of work that applies a combined guilt and shame scale (frustrating, since shame and guilt do not operate in the same way on decision-making!). I would consider shame to be the least studied of emotions with potential relevance for pro-climate action, which gives us little to work with other than conjecture, so let's proceed with some conjecturing. What is shame? Along with guilt and pride, shame functions within groups to support conformity with group norms and expectations. Guilt is an emotional reaction to a specific expectation to either take or refuse to take a certain action. For example, I feel guilty every time I fly, or don't make it out to a local climate protest. Shame is an attack on one's self-worth, a perceived poor evaluation of oneself on the part of others. The attacks that many climate advocates, myself included, endure on the part of climate deniers are intended to evoke shame, as many such attacks are in reference not to something one has done, but rather to who we are, a core feature of our identity—an activist say, or a scientist. Or a communist.

Since shame is such a terrible feeling, regardless of the merits of the shamer, it often supports withdrawal from the situation. The attacks of climate deniers in my social media feed are one reason I tend not to engage in social media as much as I might do otherwise. Its absence from the extant literature on climate attitudes and behaviours might make sense in that case. However, it can, and—speaking again from personal experience—does serve as an inducement to pro-climate behaviour, when the norms and values of one's in-group favour climate concern and action. When frequent flyers, meat eaters, and truck owners are made to feel shame in their in-group, these interactions can serve as powerful motivators supporting pro-climate behaviour change (Flight shame is starting to attract research attention, see Gössling[28]). While these anecdotes may describe a minority of group settings today, the more pro-climate action becomes normalized, the more influential shame will be in ensuring compliance with those norms.

Unlike shame, pride feels awesome, and is thus considered a strongly motivating emotion. One feels pride when one's personal identity and social standing within a group are confirmed. Pride is thus potentially effective even in the absence of climate alarm, as the main goal may well be a reinforcement of good standing in one's community, rather than challenging the climate emergency. One common route to pride is by copying what others are doing, particularly when those others have strong reputations in the group, and this is just as true for pro-social and pro-environmental behaviours as it is for conspicuous consumption.[29] As a result, it may be particularly relevant to the adoption of new practices and technologies that can be publicly displayed, like a solar panel array, or electric vehicle.[30] Empirical studies that have focused on the role of pride in pro-environmental or pro-climate behaviour often seek to compare the relative weight of pride and guilt on inducing pro-climate behaviours, and not surprisingly given the small the number of studies to date, the empirical record is mixed. Guilt appears to play a stronger role in some studies,[31] while pride is found to be a stronger precursor in others,[32] although I suspect this disagreement is the result of different methods more than anything else.

Empathy

I am going to say much more about empathy than shame and pride, because we have a stronger research record to work with, but also because I think it is possibly the most important emotion with bearing on social responses to the climate emergency. I fully agree with Birgit Müller,[33] who argues that 'for humans to engage critically with the world and become able to act politically, abstract moral law and rational thinking are necessary, but not sufficient; such engagement requires first and foremost empathy with others.' Müller goes on to support this argument by describing how farmers and gardeners from all over the world emerged to support Percy Schmeiser, a Saskatchewan farmer sued by Monsanto for patent infringement for reseeding his crop, because of a shared understanding of the injustice of such a violation of Schmeiser's autonomy as a farmer. As Thierman and Sheate[34] note, anticipated pride, shame, or guilt are forms of 'controlled motivation.' That means those actions to a great extent are governed by others, and we succumb to those social pressures because of our desire to belong. On the other hand, when we pursue a course of action on the basis of one's personal values, on the basis of the things we care about, that is a fully autonomous form of action. Social obligations can shift, personal values much less so.

Empathy was defined in Chapter 4, so in this section I want to expand upon that opening in order to explore emerging empirical research linking aspects of empathy to climate concern and action. To recap, empathy is rather complicated, consisting of three elements: *affective sharing*, which describes an autonomic response to the situation some other being is experiencing, usually negative, like distress or agony, but it could be positive as well; *empathic concern*, describing the degree to which the observer feels care and concern for what that other being is experiencing; and *perspective taking*, a cognitive capacity to put oneself in the shoes of another, which, particularly in the case of dissimilar Others, implies a deliberate choice to invest the effort required to do so. Theoretically, empathy is strongest as an action motivator when all three of these elements are acting in concert, although one can imagine several instances when a behavioural response might be induced in the presence of just one or two of these.

We discussed in Chapter 4 how empathy is most readily invoked in response to situations faced by others in our in-groups, beginning with those with whom we are directly connected—family, community, team members, classmates and co-workers—and extending to those with whom we identify based on social identities, like ethnic, occupational, political, or national identities. Such expressions of in-group empathy may in certain circumstances support pro-climate action, particularly in situations in which members of that in-group are recognized to be particularly vulnerable to the impacts of climate change. In-group empathy cuts both ways, however, shaping for example whether one experiences a fear *for* climate refugees, or a fear *of* climate refugees. For this reason, expanding our empathy maps, to use Arlie Hochschild's phrase, is an important element to pro-climate action. This is so because, first, the distribution of climate vulnerability is highly unequal, with some groups highly vulnerable, while others are relatively well-insulated. If the former group settings are the only spaces in which we can expect climate-related empathy to emerge, and not in the latter group, which happens to include a large proportion of individuals who are either playing a disproportionate role in protecting those systems associated with high emissions, hold a disproportionate level of political influence to be able to support change, or both, well then, our prospects for successful transition will be a whole lot lower.

Fortunately, empathy map expansion is something we are all capable of, and many of us have already done so in many ways, as when Global Northerners donate to an international charity, or settlers show up as allies to an Indigenous rights protest. Sara Ahmed[35] confirms our capacity to feel for others who may be remote, or distant from us. This is, Ahmed continues, a reciprocal process: as our feelings towards an Other emerge and intensify, that other appears closer in our gaze, breaking down old and establishing new borders surrounding the collectives with which we belong.

Empathy map expansion thus requires some form of wake-up call, a situation that prompts consideration of others beyond our in-group. Certain personal experiences can lead to empathy map expansion, for example, when those experiences are shared among individuals not currently in your in-groups. Some studies have evidenced empathy map expansion after direct experience with an extreme event, as survivors come to embrace more favourable attitudes towards other survivors of similar events, even when those survivors were formerly associated with one's out-groups.[36] Seo, Yang, and Laurent[37] found that exposure to cues that elicit awe can serve this purpose as well, and even elevate our empathy map expansion to the globe. In their work, these researchers showed that being exposed to such prompts can evoke a greater appreciation of our interconnectedness, which in turn motivated donations to international charities. On the flip side, shared grievances work similarly, including a shared sense of unfairness. Anandita Sabherwal and colleagues[38] found that invoking public anger regarding government inaction on the climate emergency can support enhanced pro-climate action, even across partisan divides.

Now, we've been talking a lot about *people* in this section on expanding our empathy maps. And for those Westerners who have been told that only people are sentient, valued beings and everything else is just matter, this may make perfect sense. However, for many, many humans on this planet, community means more. Community includes all those more-than-human beings, it includes the land, the planet itself. Variations on this worldview are embraced by many Indigenous peoples around the globe, and by some non-Indigenous peoples as well. The point is, if we have the capacity to call our neighbour sister, we also have the capacity to call a mountain, or a snake, our brother. This raises a whole new dimension for empathy, a *beyond-human empathy,* to operate on our emotional wellbeing, and it is a powerful route to pro-climate action. As Christopher Beck[39] writes,

> our emotional experiences with the other-than-human ... illustrate how a gestalt shift from 'humans as apart from' to 'humans as embedded within' complicates the moral picture of how we live with and in this world. I argue that when humans attend to our experiences with nature in an open and caring way, we can more easily and accurately ascertain the moral significance of the other-than-human parts of nature.

What is more, the dominance of human exceptionalism has many costs associated with it, but one big one is the cost to our own personal wellbeing. Because of course we *are* members of socio-ecological communities whether we recognize that fact or not, and the lack of opportunities to enjoy, give to, and receive the gifts from the relationships that those communities offer, create personal holes that many of us don't even know exist, until we have a chance to be reminded, through access to natural places, through caring for pets, plants and gardens. Research focused on empathy with nature is scarce, but a study by Kim-Pong Tam[40] is a valuable exception. As should come as no surprise, Tam's work confirms that empathy with nature not only generates distress about its demise, it also motivates acting *for* nature.

Complexity Thinking

My final three agency mechanisms—complexity thinking, future-scaping, and hope, are closely related. Complexity-thinking rarely comes up in discussions of pro-climate action, academic or otherwise. Some closely related concepts have been duly noted though. Bright and Eames,[41] for example, found a propensity among climate strikers to delve into contemplation of history, and the strikers' place within the biosphere—two important aspects of complexity-thinking. And quite a few studies of environmental attitudes note the importance of an awareness of consequences to environmental concern, which is also a basic tenet of complexity-thinking.[42]

I never gave complexity-thinking much thought in my own research either. I began thinking about it, however, after it emerged in one of my own studies of climate change leaders in Alberta, which I'll get to in a minute, but first, I want to clarify just what is meant by complexity-thinking. Complexity-thinking involves first an awareness of the complex nature of many systems, particularly our socio-ecological systems. I find the work of the late Charles Perrow[43] especially useful here. In his study of a variety of technological disasters, Perrow showed that those disasters erupted from within systems that were uniquely difficult to manage and control, because the systems in question were associated with high levels of interactive complexity, meaning that component parts were highly connected to each other enabling nonlinear pathways of effect, and tightly coupled, such that a change in one component can quickly generate ripple effects throughout the system. Because these qualities translate into nonlinearity, it also opens avenues for multi-causality. In one of his best-known case studies, that of the Three-Mile Island near-catastrophic accident at a nuclear power plant on the eastern seaboard of the U.S., Perrow shows compellingly that the worrying situation—a hydrogen bubble in the core—was a cumulative effect; it was the outcome of multiple occurrences, some as minor as a burned-out light bulb, and the failure to return a valve to the correct position after cleaning.

Complexity-thinking also necessarily entails clear-eyed consideration of the many uncertainties embedded within those systems, uncertainties that smother our predictive powers. This is a hard one. We all have discomfort with not-knowing, which is one very big reason why we embrace stories, seek explanations, spokespersons, or data, that provide us with that sense of restored certainty and stability, however fallacious. Complexity-thinking involves a refusal to grasp onto such stories, and sitting with uncertainty, including acknowledgement of multiple plausible change pathways, along multiple time frames from fast to slow, and including the possibility for Black Swans, those outcomes that are unimaginable until they occur.[44] To grasp, and consider approaches to overcome, complex problems such as the climate emergency, complexity-thinking is crucial. For one thing, complexity-thinking overrules the temptation to draw simplistic conclusions, to reach for those one-dimensional explanations that favour complacency, like 'everything is going to be fine' or 'we are doomed.' Complexity-thinking encourages serious consideration of all plausible outcomes, even the very dire, while also allowing for contemplation of multiple pathways of intervention to avoid those outcomes. For another thing, and most importantly to my mind, complexity-thinking places an emphasis on agency—it tells that actor that they have the potential to change the course of the future.

Back to that study of climate leaders in Alberta.[45] This study emerged from a larger project focused on evaluating climate change vulnerability in small towns in Alberta, a province in which, I might add, climate skepticism is alive and well, with the boot prints of the fossil fuel industry covering the landscape. While interviewing a wide sampling of stakeholders in a number of Alberta's small towns, I began to take a keen interest in that subset of respondents who were particularly active in pursuing responses to climate and environmental change—I called them 'climate change meta-reflexives.' These were individuals for whom such projects had become central to their identity, through career choice, and also personal and social practice. What really struck me in this study was not the additional commitments to addressing the climate emergency that these individuals expressed in their practices, laudable as they were. Instead, I noticed the distinct ways that these climate meta-reflexives conceived of the problem, and how to address it, in comparison to other interview participants. While most interviewees could identify one or two anticipated impacts of climate change, usually reflective of their personal interests, such as a decline in opportunities to ski in Winter, the climate meta-reflexives went to great lengths to articulate the multiple, interactive, and systemic impacts posed by global warming. They included in these accounts not just local, and more or less short-term impacts, but global, and long-term impacts, including contemplation of potentially catastrophic outcomes. Consideration of worst-case scenarios, in other words. Rather than leading to despair, however, these individuals were motivated to act. This could be attributed to their high ascription to norms of responsibility, but their interviews also indicated that they were convinced that even seemingly small acts—planting a community garden or organizing a community discussion, for example—mattered. In other words, they had confidence in their own agency, and our collective capacity to institute change, which seemed to emanate from their complexity thinking. This motivation to act, moreover, appeared to emanate from an acceptance of the uncertainty involved. Whereas other interviewees referred to uncertainty as a reason not to act, the meta-reflexives came to the opposite conclusion.

Future-gazing

Complexity-thinking and future-gazing tend to go hand in hand. Awareness of consequences is an expression of future-gazing, but there is much more involved. There is an inevitable time dimension that comes to the fore with the application of complexity thinking, which encourages not only consideration of history, but also of multiple possible futures. Those climate meta-reflexives were unique in their complexity-thinking, and also in their future-gazing. Whereas many other interviewees expressed rather limited levels of future-gazing, resorting rather quickly to deterministic future prognoses: either we are already doomed, or human ingenuity and technology will save us, the meta-reflexives expressed deep concern about, and contemplation of, the *possible* futures that may unfold-in other words, motivated by maybe. One climate meta-reflexive, Jane, (a pseudonym), for example, ponders,

> If we actually could make this transition in a way where we're not actually starving yet, but we still can feed ourselves ... whereby there's these crises that happen that are gonna wake us up to do these things that we really should have done fifty years ago, but that we do them in a peaceful way and without the crisis being so big that we're actually losing our homes or whatever. ... That would actually be quite exciting.

Exciting indeed! Thinking about the future *can* be exciting, exhilarating even. It can also be very worrying. Future-gazing invites in scenarios of apocalypse, utopia, and everything in between. Many future narratives are culturally prescribed. Progressivism, a centre-piece of Western worldviews, offers a lovely vision of the continued march of social progress, towards ever richer, longer-lived, and more leisurely lifestyles (lovely anyway, for the white, the privileged, the male, the settler, the middle class; less so for the many upon whose lands and backs pursuit of this vision was wrested). Alongside this vision, Western science cultivated a confidence that the future can be predicted and controlled, via science, technology and risk management, and therefore, this progressivist future was not merely a *possibility,* it was an *inevitability.* So inevitable, in fact, that one not need raise a finger to ensure it comes to pass. The assurances offered by this myth in many ways, in other words, deter deliberative future-gazing, eliminate the need to grapple with uncertainty, and motivate complacency, or at least limits one's actions to self-improvement to make sure you get a slice of that pie.

So, many of us learn to approach the future in certain ways from the cultural structures within which we are embedded. But future-*gazing*—deliberating, contemplating—is indelibly an aspect of personal reflexivity; an inclination to question culturally prescribed narratives, and imagine alternatives. The arts and popular media are replete with future imaginaries. While climate dialogues are filled with visions of doom, prospects of apocalypse are not new. In the modern era, imaginations of a nuclear winter were deeply embedded in popular discourse during the cold war, and apocalyptic visions have been a mainstay of the modern environmental movement.[46] But there is something different about the climate apocalypse imaginary. For one, it is not simply a possibility, there are signs of it already unfolding. For another, the danger does not rest with a small handful of actors with access to the 'red button'; the danger is imposed by systemically embedded elements of modern society itself.

Social movements need shared visions of the future, in order to construct collective identities. Environmental movement organizations often grapple with finding the right future gaze to motivate support for social change. Cåssegard and Thörn[47] note that the environmental movement is unique among social movements in its adoption of a dystopian, apocalyptic future imaginary—a future to be avoided—in comparison to most modern social movements that instead invoke a utopian future (equality, say) to be pursued. Movement organizations' attempts to grapple with the climate emergency have broadened the landscape of future imaginaries, however. Some mainstream organizations, for example, have concluded that apocalyptic imagery, perhaps because it feels a bit too real, might be more likely to generate despair, rather than activism, and have opted to rally around more positive future imaginaries instead. Not so for some of the more 'radical flank' organizations, which have drawn a different conclusion, arguing that citizens need a wake-up call. As Cåssegard and Thörn observe, however, a third imaginary has surfaced, particularly among, and with respect to, social movement organizing in the Global South. In this narrative, the apocalypse is already here. This post-apocalyptic narrative draws attention to losses that have already occurred, and future losses that are at this point unavoidable. Here fear remains, but perhaps even more prominent is anger at the blatant injustices that have precipitated this apocalypse. Movement goals include accountability, in addition to avoidance of further harm.

This splintering of narratives across today's climate movements is telling, a reminder to apply the term 'We' carefully. Note the huge erasure that happens when we speak of a *global* future, erasure of the stark differences in the pace of apocalypse unfolding between North and South. Among the middle classes of the Global North, emerging extreme events have come as a shock, just becoming visible in our peripheral vision amidst a background in which such events remain a departure from normality. In the South, and even in the Fourth World of the Global North, such as many Indigenous communities in northern Canada, what is for middle class residents scarcely imaginable has become an everyday reality.

Even post-apocalyptic narratives involve future-gazing though, a future in which justice is served. And while there are by no means guarantees that the future imaginaries pursued through collective action will be realized, it is safe to say that the probability of their realization, even in part, or in variation, is increased the more such imaginaries are shared, and acted upon. By imagining the future, we are, in effect, making it. As Beck and colleagues[48] argue, the creation of shared catastrophist narratives of the future coincide with cognitive, affective, and political preparedness to confront those futures. Clot-Garrell[49] nicely captures an extensive field of sociological research that attends to the significance of future-gazing:

> future imagination [is] not merely a mental representation and discursive construction, but a driver of action. Futures are 'productive' since they affect people's present decisions, practices and relations, but they are also 'produced' in the way these shape latent futures of our making. These latent futures also constitute reality as they 'live within the present', though they are still not fully developed or manifested.

So, future-gazing, and the emotions it invokes, are important. Dreamy, utopian thinking is absolutely important, particularly for youth, but it is perhaps even more important not to look away from potentially very dark futures, even if their probability seems small, order to prepare for and possibly avoid their realization. Future-gazing is far from inevitable, however, and under certain conditions it may be scarce. I mentioned earlier that certain strong prescriptions of the future can shut down future-gazing, when those futures are offered with deterministic, certain, proclamations. There is another route to shutting down our future-gazing, however. How can we even imagine alternative futures if we feel so completely disempowered, or don't even have the mental space to invest in the reflexivity required, when all our mental energy is required simply to make it to the next paycheck? Or, when positive future pathways simply appear unimaginable. The anxiety epidemic among youth we are observing today may be at least in part associated with the inability to envision a future worth living.

Hope

Like fear, hope is anticipatory, it is speculation about the future, replete with uncertainties. Hope is also unlike fear in one very important respect. Fear is an autonomic, biophysical reaction. In some instances, our bodies detect fear before our minds even catch on, producing that tingle, that raising of hair at the back of your neck. Unlike fear, hope requires contemplation—it is a feeling, in other words, that demands full engagement of our frontal lobes. In addition, while fear is most acute in response to an immediate, tangible threat, hope has everything to do with future-gazing. To embrace hope thus also reflects an engagement with uncertainty. It is about potentialities, plausibilities, much more than probabilities. Hope requires attention to the future, an orientation towards time, an as yet intangible, not-yet-reality, although with real consequences for our decision-making and hence, our future realities. Hope is felt today, but the vision or object that inspires the hope has not yet come to pass, and may not. Indeed, the very concept of hope entails awareness of the possibility that the object of hope may never come to pass.

A big mistake many people make is to confuse hope with optimism. In many ways, however, hope is optimism's Janus face. Optimism implies a probability estimate. If I am optimistic about a particular outcome coming to pass, it means I believe the probability of that outcome coming to pass is higher than not. Hope is something else entirely. I *hope* when probabilities are small, perhaps even vanishingly small. Emerging research focused on hope belies the conclusions of earlier research indicating that we are most motivated to act when we expect success.[50] This conclusion is not necessarily wrong—we do often act when we perceive the odds of success to be high; we also may act on the basis of a rational calculus of our personal interests. But, as discussed, okay harped about, throughout this book, we are not *solely* rational utilitarian actors. We are much more than this, and recent research that moves us beyond the rational-actor assumptions of much previous human behavioural research has been enormously revealing in this regard. This includes the latest studies of social responses to climate change, which show that hope is a particularly strong motivator for action in comparison to other stimuli,[51] and it actually performs better than optimism as a motivator for taking up various forms of pro-climate action.[52] Optimism, after all, can support complacency, a presumption that I am not needed to fix this problem. Technological optimism is a particular form of optimism associated with Western worldviews that contributes to complacency regarding environmental and climate issues.

What gives rise to hope then? And how can we remain hopeful even when the indications of failure just seem to continue to stack up? Why might I be inclined to assume such a seemingly irrational position? We certainly do not assign hope to every situation that entails slim chances of success. I hope, despite low odds of success, when the outcome in question is one that I care about deeply.[53] When my sister was diagnosed with stage four melanoma, I refused to lose hope that she would make it through, despite the low statistical probabilities of survival (fortunately, as of this writing, she has beat the odds!). Hope mediates our fears, and enables us to take action despite them.[54] Perhaps ironically, since hope is associated with low odds of success, it has also been shown to enhance perceived efficacy.[55]

As likely comes as no surprise, hope plays a central role in social movements, including climate movements.[56] Hope both motivates engagement, and sustains that engagement, particularly as it becomes a collective emotional experience. Hope also inspires creativity in collective action.[57] That creativity is particularly important in the face of the climate emergency, for which the object of hope is much more elusive than a loved one facing a terminal illness. In this situation, the activation of future imaginaries comes to the fore. Without a vision, even a utopian one, of possible futures, what is there for hope to attach itself to?

Hope is slippery, though. As noted by Paulo Friere, in his book, *Pedagogy of Hope*[58] (2014), hope must be fed by action, or in Freire's words, 'anchored in practice.' If we don't act on hope once it surfaces, it will quickly slip away.

Conclusions

Inaction is always easier than action, but action is always necessary. We all take action every day to do the things we love, like hiking or playing soccer of course, but we also take action to accomplish other things we would rather not do, like preparing meals, getting to work, cleaning the house, exercising, and helping out those in need. All forms of pro-climate action also take a high level of effort, to varying degrees, but boy are they necessary. Fortunately, there are multiple intervention points that can stimulate a shift from inaction to action, and identifying these intervention points can go a long way to support strategic efforts to upscale pro-climate action in its multiple forms.

I'll dig into some intervention ideas more in the next chapter, but they are informed by two key arguments made here. First, alarm triggers must be felt as well as heard. The messages that our warming planet convey will just be whispers in the wind unless they are attached to something of value. Even then, many of us may still not hear the warnings, because of all the background noise in our lives. Those who hear the warnings are, first of all, listening, because they trust the source, whether a friend or a newscaster, and second, they have the mental energy to apply meaning to those warnings; to perceive what those warnings imply for social and ecological wellbeing.

Warnings heard are not always heeded, however. Agency must be activated by one or more of the seven mechanisms discussed here: norms of responsibility; efficacy; belonging; empathy; complexity thinking; future-gazing; and hope. At least one of these must be strongly present, and none strongly absent, to support pro-climate action, and the more that are present, the greater the level of motivation.

Pathways to Action, or Doing the Hard Thing   151

Action takes work, but it also offers many rewards. People who exercise know this—getting out the door for that run always takes a bit of self-talk, but then you're oh-so satisfied afterwards. This is true when we take action to support social change as well, and this is particularly the case for those of us who are inclined to feel despair. The act of *doing* itself becomes its own form of therapy.[59] But the effects of overcoming inaction run even deeper than this. The very process of engagement in pro-climate action, particularly collective forms of action, can enhance those mechanisms of agency, generating a positive feedback loop. Action begets action as it were, by, for example, offering opportunities to evidence efficacy,[60] counteracting guilt, or experiencing the contagious spread of hope. The linchpin in these positive feedbacks is the opportunity for positive emotional rewards that can keep us afloat despite the weight of our fear and anxiety, our guilt, and the shame and antagonism many pro-climate actors face. These positive emotional rewards are essential to the sustained pro-climate action needed to support transformational social change.

Notes

1 Olson, *The Logic of Collective Action*.
2 Davidson et al., "Just Don't Call It Climate Change."
3 Peters and Slovic, "The Springs of Action."
4 Bradley, Cuthbert, and Lang, "Picture Media and Emotion"; Shoemaker, "Hardwired for News."
5 Diamond and Urbanski, "The Impact of Message Valence on Climate Change Attitudes"; Bloodhart, Swim, and Dicicco, ""Be Worried, Be VERY Worried"; Chapman et al., "Climate Visuals"; Witte and Allen, "A Meta-Analysis of Fear Appeals: Implications for Effective Public Health Campaigns."
6 Bright and Eames, "From Apathy through Anxiety to Action."
7 Wang et al., "Emotions Predict Policy Support."
8 Bouman et al., "When Worry about Climate Change Leads to Climate Action"; Wang et al., "Emotions Predict Policy Support"; Galway et al., "What Drives Climate Action in Canada's Provincial North?"
9 Wang et al., "Emotions Predict Policy Support"; Van Boven et al., "Feeling Close."
10 Conte, Brosch, and Hahnel, "Initial Evidence for a Systematic Link between Core Values and Emotional Experiences in Environmental Situations."
11 Swim et al., "OK Boomer."
12 Stollberg and Jonas, "Existential Threat as a Challenge for Individual and Collective Engagement"; Bouman et al., "When Worry about Climate Change Leads to Climate Action."
13 Bouman et al., "When Worry about Climate Change Leads to Climate Action."
14 Spence, Poortinga, and Pidgeon, "The Psychological Distance of Climate Change."
15 Hurst and Sintov, "Guilt Consistently Motivates Pro-Environmental Outcomes While Pride Depends on Context"; Adams, Hurst, and Sintov, "Experienced Guilt, but Not Pride, Mediates the Effect of Feedback on pro-Environmental Behavior"; Lu and Schuldt, "Exploring the Role of Incidental Emotions in Support for Climate Change Policy."
16 Harth, "Affect, (Group-Based) Emotions, and Climate Change Action."
17 Wolsko, Ariceaga, and Seiden, "Red, White, and Blue Enough to Be Green."

152 Pathways to Action, or Doing the Hard Thing

18 Clark and Adams, "Altering Attitudes on Climate Change."
19 Bostrom, Hayes, and Crosman, "Efficacy, Action, and Support for Reducing Climate Change Risks."
20 Toivonen, "Themes of Climate Change Agency"; Chen, "Self-Efficacy or Collective Efficacy within the Cognitive Theory of Stress Model"; Fritsche et al., "A Social Identity Model of Pro-Environmental Action (SIMPEA)."
21 Wang, Geng, and Rodríguez-Casallas, "How and When Higher Climate Change Risk Perception Promotes Less Climate Change Inaction"; Gregersen et al., "Outcome Expectancies Moderate the Association between Worry about Climate Change and Personal Energy-Saving Behaviors"; Bieniek-Tobasco et al., "Communicating Climate Change through Documentary Film."
22 E.g. Traister, *Good and Mad: The Revolutionary Power of Women's Anger*; Flowers, "Refusal to Forgive:Indigenous Women's Love and Rage."
23 LeBrón, "Policing Coraje in the Colony: Toward a Decolonial Feminist Politics of Rage in Puerto Rico."
24 Letourneau and Davidson, "Farmer Identities."
25 Jenkins, "Why Do All Our Feelings about Politics Matter?" P. 201.
26 Karsgaard and Davidson, "Must We Wait for Youth to Speak out before We Listen?"; See also Zaremba et al., "A Wise Person Plants a Tree a Day before the End of the World."
27 Yeager, "Collective Shame in Climate Denial."
28 Gössling, Humpe, and Bausch, "Does 'Flight Shame' Affect Social Norms?"
29 Dessí and Monin, "'Noblesse Oblige? Moral Identity and Prosocial Behavior in the Face of Selfishness.'"
30 Harth, "Affect, (Group-Based) Emotions, and Climate Change Action."
31 Hurst and Sintov, "Guilt Consistently Motivates Pro-Environmental Outcomes While Pride Depends on Context"; Adams, Hurst, and Sintov, "Experienced Guilt, but Not Pride, Mediates the Effect of Feedback on pro-Environmental Behavior."
32 Schneider et al., "The Influence of Anticipated Pride and Guilt on Pro-Environmental Decision Making"; Pasca, "Pride and Guilt as Mediators in the Relationship between Connection to Nature and Pro-Environmental Intention"; Shipley and Van Riper, "Pride and Guilt Predict Pro-Environmental Behavior."
33 Müller, "'To Act upon One's Time …' From the Impulse to Resist to Global Political Strategy." P. 61.
34 Thiermann and Sheate, "Motivating Individuals for Social Transition."
35 Ahmed, "Collective Feelings: Or, the Impressions Left by Others."
36 Vezzali et al., "Feeling like a Group after a Natural Disaster"; Vollhardt and Staub, "Inclusive Altruism Born of Suffering."
37 Seo, Yang, and Laurent, "No One Is an Island."
38 Sabherwal, Pearson, and Sparkman, "Anger Consensus Messaging Can Enhance Expectations for Collective Action and Support for Climate Mitigation."
39 Beck, "Attending to the Full Moral Landscape." P. 43.
40 Tam, "Dispositional Empathy with Nature."
41 Bright and Eames, "From Apathy through Anxiety to Action."
42 Hansla et al., "The Relationships between Awareness of Consequences, Environmental Concern, and Value Orientations"; Stern, "Toward a Coherent Theory of Environmentally Significant Behavior."
43 Perrow, *Normal Accidents*.
44 Taleb, *The Black Swan*.
45 Davidson, "Analysing Responses to Climate Change through the Lens of Reflexivity."
46 Cassegård and Thörn, "Toward a Postapocalyptic Environmentalism?"
47 Cassegård and Thörn.

48 Beck et al., "Cosmopolitan Communities of Climate Risk."
49 Clot-Garrell, "Voices of Emergency."
50 Bandura, *Self-Efficacy*; Carver, Scheier, and Segerstrom, "Optimism"; Bury, Wenzel, and Woodyatt, "Confusing Hope and Optimism When Prospects Are Good."
51 Geiger et al., "How Do I Feel When I Think about Taking Action?"
52 Bury, Wenzel, and Woodyatt, "Confusing Hope and Optimism When Prospects Are Good"; Dean and Wilson, "Relationships between Hope, Optimism, and Conservation Engagement."
53 Miceli and Castelfranchi, "Hope"; Bury, Wenzel, and Woodyatt, "Confusing Hope and Optimism When Prospects Are Good"; Bury, Wenzel, and Woodyatt, "Against the Odds."
54 Kleres and Wettergren, "Fear, Hope, Anger, and Guilt in Climate Activism."
55 Cohen-Chen and Van Zomeren, "Yes We Can?"
56 Greenaway et al., "Feeling Hopeful Inspires Support for Social Change"; Jenkins, "Why Do All Our Feelings about Politics Matter?"
57 Jenkins, "Why Do All Our Feelings about Politics Matter?"
58 Freire, *Pedagogy of Hope*.
59 Schneider and Van Der Linden, "An Emotional Road to Sustainability."
60 Mortreux et al., "Reducing Personal Climate Anxiety Is Key to Adaptation."

8 Threading the Needle from Emotions to Transformational Social Change

> Those [species] which do not change will become extinct.
> Charles Darwin, Origin of Species, p. 319.

If Darwin's proposition is true for any species on earth, so too must it be true for *Homo sapiens*. Darwin has been immortalized for a very different quote—'survival of the fittest'—which has been subject to all manner of interpolation beyond the author's intent; Darwin understood well the crucial role of cooperation in the adaptation (i.e. survival) of humans, and many other species as well. 'Survival of the fittest' may well describe the competition for mates, as anyone who has dipped their toes into online dating can attest, but it most certainly does not describe an adaptive response to large, complex, existential threats like global warming.

Never before in recent human history has such a truth become so germane. The prospect for reaching the end of the evolutionary line for *Homo sapiens* is no longer restricted to the science fiction section of the bookshop. *Homo sapiens* endured some modest beginnings, with several moments at which such an outcome was improbable. For at least the last several thousand years, however, with the exception of a few scary moments (the plague, the Little Ice Age, the nuclear threat), it seemed we were here to stay. This is no longer the case. The accumulation of evidence provided by climate scientists indicates that it is by no means 'alarmist' to conclude that the rapid changes to our planetary and social systems induced by global warming, when placed on top of numerous other maladaptive trends brought into being by those in power, most certainly threaten our wellbeing, the lives of millions, and *perhaps* also our very existence as a species. Nonetheless, despite full awareness for at least 50 years of the threats posed, we've barely crossed the starting line in this race for our lives, livelihoods, and everything we hold dear.

But it's not over till it's over, as they say. This book began with the premise that, based on what we know about human behaviour and social change, we have the capacity to confront the climate emergency, personally and collectively, in ways that minimize the human and ecological toll, and maximize prospects for social and ecological wellbeing. That capacity hinges on our individual—and by extension collective—agency; and that agency is invoked, or not, along pathways that are fuelled by our emotionality. Our fears, our passions, our hopes and desires have always served as our compass, and this remains as true today as it was for our ancestors. We must feel our way to change as much as we think it. Foremost, alarms must be triggered at an emotional level. The climate emergency must be received as a violation of our senses, a violation that evokes fear and anxiety in accordance with the existential threat that is posed, and rage at the gross injustices embedded in both the causes and consequences of our rapidly warming planet. Second, however, taking action cannot be *solely* a labour of fear, or of anger. It must also be a labour of love—the passion, and compassion, that motivates us to encircle our protective arms around the beings and things we care about.

What makes our emotional compass tick? What are the magnetic poles at work? Our emotional compass operates to support the basic human needs for thriving. Meeting those human needs, in turn, enables agency, and allows for the manifestation of those human qualities that are necessary for collective problem-solving: in other words, our collective survival. Many scholars have developed extensive lists of what are considered basic human needs, but for our purposes of facilitating transformational social change, I think we can zero in on three:

1) *Reflexivity*: the mental capacity to be able to problematize one's circumstances and evaluate their causes, consequences, and points of intervention; access to sufficient knowledge and information to do so.

2) *Efficacy*: the confidence that one can make a difference, and access to the rights and resources required; being the protagonist in one's own story.
3) *Belonging*: having one's identity nurtured and validated by others; having social spaces in which to share in creating culture, and receive emotional support.

The four inaction pathways described in Chapter 6 each have unique fingerprints, but all describe personal reactions to compromised reflexivity, efficacy or belonging in one way or another; conditions that evoke emotional responses that serve to support inaction. Apathy describes, more than anything, compromised reflexivity, perhaps instilled by belonging to a group in which apathy is reinforced. Denial for many emanates from threatened belonging—strong affiliation with a group whose identities, beliefs, and practices have been implicated—and the defensive postures that result in turn delimit reflexivity, motivating a rejection of new information that could support learning. Many deniers are also situated within conservative and/or rural social milieus, which celebrate independence, a 'frontier' survivorhood. That milieu prescribes high levels of personal autonomy, enhanced by deep distrust of institutions or other bodies that are perceived to be threatening that autonomy, particularly government, leading to reactions like 'I have to see things for myself,' and 'don't tell me what to do.' To accept sight unseen the conclusions of government representatives or climate scientists is anathema.

Withdrawal describes a lack of efficacy, which prevents alarmed individuals from responding to their anxiety. This situation imposes substantial compromises to personal mental health, which in turn hobbles one's reflexivity, fostering information avoidance. Those experiencing high levels of anxiety may also be inclined to resort to withdrawal due to the absence of emotional support, in other words, belonging. And being stuck describes conditions in which reflexivity may be present, but what is missing is efficacy, reinforced by belonging to a group that does not provide peer support for taking action.

The presence of all three basic human needs—the investment of reflexivity into personal contemplation of the causes, consequences, and potential routes of action; confidence in one's personal efficacy as well as the efficacy of collective action; and belonging to a group that encourages action—sets the stage for committed personal and collective action with the intention of addressing the climate emergency. But action pathways are likely to be tenuous, even for those committed to taking action today, without concerted attention to the continued revitalization of those three prerequisites.

Upscaling and sustaining climate action at the level necessary to support rapid, transformational social change thus demands that those currently committed to addressing the climate emergency turn their attention to confronting those social structures and processes that compromise reflexivity, efficacy, and belonging. The number and variety of personal and collective projects that have the potential to enrich one or more of these elements are only limited by the imagination and creativity of their agents and may not look like *climate* action at all. A community garden can cultivate efficacy and belonging; a stitch-and-bitch club can invite reflexivity; supporting voter rights and poverty reduction builds efficacy. Working to reconcile Indigenous-settler relations, beginning with we settlers according due respect for Indigenous rights, knowledge, and ways of knowing, can build upon all three. In this closing chapter, I will focus on four key priorities for disrupting inaction pathways and sustaining and upscaling action pathways by seeking to create conditions for reflexivity, efficacy, and belonging: cultivating empathy; envisioning futures worth fighting for; bottom-up politics; and feeling our way to change.

Getting to We: Cultivating Empathy

To be alive is to be in relation with others, human and non-human alike. Our ability to survive depends upon attending to those relations to ensure reciprocity, the key ingredient for cooperation. We do so with empathy. Empathy begets reciprocity begets cooperation. Empathy reminds us that in order to receive—food, respect, love, healing—we must give. But empathy does much more; it provides us with the *desire* to give, even without the expectation of reward. When I help an elderly person in distress, I do so not because I have any expectation that she will be in a position to help me tomorrow, but rather because the act of helping itself is emotionally rewarding to me. We are each born with an inherent tendency to feel what others feel, to care about those others, to expect fairness for ourselves and for others. We are also born with the cognitive capacity to 'walk in another's shoes,' although the tendency to so do might require some social learning. Empathy is most readily expressed within our immediate families, our clans and communities, because these are the communities within which we engage in relations every day, relations that enable us each to survive, and hopefully, to thrive, but we are also capable of expanding our empathy maps to include distant others.

There are other members of these communities who are just as important to our survival and wellbeing. The nonhuman beings around us, the land beneath our feet, the rivers, the wind, the storms, these are all parts of our communities as well, with whom we interact on a daily basis and from whom we have taken much. Many peoples enjoy reciprocal, empathic relations with these nonhuman members of their communities, but we among Western cultures, those embracing capitalist, patriarchal, and colonial worldviews, have lost sight of our interdependent relations with the nonhuman members of our communities, and with non-Western peoples as well for that matter. Our disregard for these relationships has led to a breakdown in reciprocity, to alienation within our socio-ecological communities. Is it any wonder we settlers never seem to feel at home anywhere?

The climate emergency is emblematic of this disregard, and thus addressing the climate emergency demands reconciliation of our socio-ecological relations. As in all relationships, when one partner is doing all the giving and the other doing all the taking, that relationship is eventually going to break down. We among Western cultures need to acknowledge that 'we' includes the nonhuman members of our communities, to whom we owe a massive debt. Cultivating empathy with nonhuman community members requires that we first acknowledge that we are not exempt, we are not exceptional, we do not have privileged status among other members of our socioecological communities that magically allow us to disregard our interdependence. It also demands that we relate to nature not as an object, but as a subject, animated with agency, rights, feelings, as members of our communities equal to ourselves.

The transformational social change needed to confront the climate emergency also requires the expansion of our empathy maps beyond our immediate communities. While these extra-local relations are less immediately available to our senses, we also engage with nonhumans, landscapes, waterways, and planetary systems that are far beyond our immediate surroundings, relationships recognized by many non-Western worldviews, and brought to light by climate and ecological scientists as well.

We also interact with, and are indeed interdependent upon, non-local peoples, with whom we now must find ways to cooperate. Cultivating empathy for non-local peoples to enable such cooperation requires that we reflect upon who is 'we.' Because empathy is most readily expressed in our immediate families, clans and communities, this is the 'we' to whom we tend to refer. Group living has allowed for the cultivation of a rich diversity in cultures, beliefs, and knowledges across the globe, which have become defining features of who 'we' are, generating multiple 'we's'. And, given that our current social systems are compartmentalized hierarchically, with power, agency, and worth assigned to a small minority, the intersectional position of groups also becomes a critically important dimension to the word 'we.' Cultivating an expansive empathy requires that each of us honours and preserves our individual and group uniqueness—retaining this cultural diversity is critically important to addressing complex problems—while also confronting identity-based oppression and inequity in our politics by bringing those intersectional positionalities into sharp relief. Doing so is crucial to the pursuit of justice, without which reciprocity is impossible.

Enabling the cooperation required for transformational social change demands justice; but the pursuit of justice, while absolutely necessary, is not sufficient; it must also include cultivating empathy across those group boundaries. Cooperation is not the only thing that is lost in a divided world. The more divided we are, the more preoccupied we are with the things that divide us rather than the things we hold in common, and the more dangerous it is to the social belonging of any individual to critique an idea that emerges from their group, or explore the potential merits of the position taken by another group. In fact, in a starkly tribalized social universe, there is little room for reflexivity at all; only fear and distrust prevail. While we remain divided, the only means of engagement is conflict.

It doesn't have to be this way. Indeed, it must not be this way if we are to confront the climate emergency. We can honour our unique identities while also expanding our empathy maps to meet others on level ground, in safe spaces, among whom we practice reciprocity. The more we talk to, and even more importantly listen to, strangers, the less dissimilar to us—the more human—they become. Conflict is an inevitable feature of group living; however, addressing the climate emergency requires that we move beyond conflict towards cooperation, and cooperation can only happen in the presence of empathy; it requires that alongside our differences we forefront our commonalities. We are all already members of the same community, metabolically dependent upon the same planet; we are also all intelligent, feeling mothers, daughters, partners, friends, teachers, learners; and we all seek to feel capable, to belong.

Start Where You Are: Bottom-up Politics

To act as if politics cannot change the moral and behavioural outlook of citizens, by identifying their more generous and relational human spirit and giving collective articulation, hope and public presence—and material policy form—to these elements of human nature, is timid, wrong, and maybe even self defeating.

Brian Wynne, 2010

Democracy has taken a beating of late, with democratic indicators on the decline across the world, and the rise to power of right-wing leaders with autocratic ambitions on several continents. The health of little-d democracy, I would argue, has been under duress for far longer. Even some career leftists have begun to question the utility of democratic systems of governance for addressing large-scale, long-term crises like global warming. Authoritarianism just seems so much more efficient, avoiding the party conflicts, stalemates, policy compromises, and short-termism dictated by electoral cycles. Imagine a benevolent, ecologically wise dictator who can punish polluters and roll out green carpets with the snap of her fingers!

A green authoritarian state, however, is just a fantasy; as with so many other quick fixes, it is too good to be true. One need only consider the types of people who tend to seek positions of authoritarian rule to give one serious pause. Empaths do not become dictators. Dictatorships also do nothing to challenge the entrenched power structures that serve to empower elites and disempower everyone else; the very power structures that are a main reason we are in this mess in the first place. To the contrary, dictatorships are designed to protect those power structures. But democracies do indeed appear to be frustratingly shackled when it comes to confronting environmental and climate crises, with the few wins, in the form of national parks and air pollution regulations to take a couple notable examples, mere exceptions that appear to prove the rule: democratic systems of governance have done nothing to prevent those with power and privilege to blast through global resources and spew waste at an ever-escalating pace.

While Western, representative democracies are indeed faring poorly in their handling of the climate emergency, I believe that the lousy track record is a reflection of the poor health status of little-d democracy within those systems and not an indication of the superiority of authoritarian forms of governance. As every high school civics student in the West will likely be able to recite, democracies are essential to the protection of individual liberties. But far less discussed is the fact that the realization of those liberties requires the commitment on the part of those rights holders to collective responsibilities. Then all we have to do is accord corporations the status of citizens with 'rights' but few responsibilities, and efforts to address social and environmental crises become little more than charades.

Democracy takes work. Authoritarianism is attractive, in part, because it does not. Maintaining a healthy democracy is like committing to a regular exercise programme, while authoritarianism permits one to sit on the couch and eat the potato chips doled out by dictators. One needs simply to follow the rule book without putting any energy into making the rules. The problem is, too many of us have migrated to the couch in our democracies too, with the active encouragement of elites in progressive and conservative parties alike, who seek the quiescent masses of a dictatorship in order to ensure the continued flows of wealth and privilege to the few, while still enjoying the legitimacy that a cloak of democracy provides.

Little-d democracy takes a *lot* of work: it requires deliberation, conflict mediation, and interacting with people who see the world differently than you. But, as with eating healthy foods and regular exercise, good skills and habits are acquired in the process. Democracy is thus more capable of supporting human wellbeing than the potato chips of authoritarianism. In other words, democracy creates space for, indeed demands of each of us, reflexivity—a basic need alongside efficacy and belonging. In fact, democracy is far more compatible with the emotional and cognitive traits that we have inherited to enhance our wellbeing than is authoritarianism, which makes sense from a historical perspective. For much of human history, we have lived in small groups that were not organized hierarchically, and it is in these contexts that we acquired our cooperative problem-solving skills. Democracy supports cooperative problem-solving by engendering reflexivity, but by bringing us together to deliberate, to contribute, to hear and be heard, democracy also engenders efficacy and belonging. Democracy, in turn, incentivizes compliance, without the need for strong-armed state control, because through engagement we all become vested interests in the outcome.

As such, democracy is far better suited to address collective action problems, including large and complex problems like environmental and climate emergencies, than is authoritarianism. And democracy happens in community, not in national capitals. Even global crises are rooted in local decisions and practices, after all. Too often, we presume that large-scale crises require large-scale—a.k.a. top-down—action plans, and in the absence of those action plans, many of us feel despair creeping in. How can we rest our hopes on an institution like the Conferences of the Parties to the UNFCCC, when that institution has little to show for 30 years of expensive, carbon-consuming efforts? How can we rest our hopes on federal policies, which are dismantled as fast as they are created, at the whims of the party in power? But there's more: even in the highly unlikely event that a global action plan surfaced, it would be doomed to fail without the local buy-in required for compliance, in homes, on farms, in the schools, stores and factories that make up our communities. If something like global cooperation is going to happen at all, it is far more likely to emerge from the bottom, in the form of multiple, simultaneous but complementary sub-global initiatives that cultivate agency among that vast majority of people who are concerned, and that gradually isolates the obstructors.

Focusing on the local, regional and group levels—where little-d democracy happens—is not only more palatable and more motivating; it is also where macro-social change happens. Through seeking to revitalize our democracies on the ground, we are laying the foundations for transformative social change to address the climate emergency. These are the spaces in which the full richness of values and priorities of local peoples can shape decision-making. A space in which we can manifest a *Vitalist Politics*, Amitav Ghosh's term for a politics in which protecting all our relatives, human and non-human, can be realized; a politics that underscores sacredness and kinship; a politics of empathy. What is needed? Collective framing of the problem, integration of our numerous unique sources of knowledge, attributes and capacities that can be invoked to address the problem creatively, and the generation of a sense of mutual responsibility and commitment. Any collective project, no matter the size or the context, that seeks to address these needs is contributing to the broader goal of transformative social change to address the climate emergency. By contrast, a top-down, command and control approach will face serious limits with respect to all three of these requirements.

Create Futures to Move towards, Not Away From

Climate action must be action *for*, not against; *towards*, not away from. It is not enough to say 'we don't want that!' We must say 'we want this!' The threats posed by global warming are dark, ugly, frightening, and deeply disempowering for many. Others, mainly voices from the Global South, have been quick to point out that the apocalypse is already here. All too many people are already living under conditions of extreme scarcity; of surviving, not thriving, and these individuals are also bearing the brunt of the climate emergency, today, not 20 years from now. The climate movement is justifiably motivated in large part by an urgent desire to raise awareness of, and seek ways to avoid, a very dark and dystopian future. Strategies to avoid this outcome posed by climate advocates, even when conveyed in the most optimistic of terms, imply costs, compromises, and actions that take the guise of general declines in quality of life.

The costs of shifting our energy and food systems, and entire economies in order to address the climate emergency are real, and should not be washed away with rose-coloured glasses, particularly given the inevitable inequities in the distribution of those costs without concerted attention to justice in our transition plans. Nor should we gloss over the inevitable impacts that have already begun to materialize, and will continue to do so for the foreseeable future. And we should definitely look squarely upon those scenarios of collapse that will come to pass if we continue along the greenhouse gas-intensive road we are on. Confronting the climate emergency necessitates clear-eyed awareness, a refusal to look away.

But for those with the good fortune to have enjoyed the fruits of our fossil fuel enriched economies, who share some of the accountability and hence responsibility to make changes, such a road may well represent a high road in comparison to the alternative, but it does little to inspire action. Simply identifying paths to avoid is no way to embark upon a journey. Many people, understandably, will hesitate to start a journey when they don't know where they are headed. Supporting cooperation and sustaining climate action requires that agents have something to strive *for;* a vision of the future that draws support, rather than solely one that inspires fear. To begin with, there are many, many positive benefits of climate mitigation and adaptation that could be emphasized in our climate narratives to a much greater extent than they are today. Take, for example, declines in air pollution from coal-fired power plants, the role that expanded urban green spaces can play in improving quality of life, the attractiveness of walkable communities, and the potential for rural communities to be revitalized with a shift to regenerative agriculture practices—which can create more jobs and higher-valued products than conventional agriculture.

But creating futures to move towards requires more than underscoring the positive benefits that counterbalance the costs of pro-climate action. Creating futures to move towards requires visioning. It requires reflexive deliberation within our communities over the needs and values that we would like to see flourish in our futures. And it requires future visions in which everyone can see a place for themselves, including farmers, home makers, city councillors, coal miners and truck drivers, and especially youth—our sons and daughters. Even when infused with a dose of utopia, the very process of creating positive futurescapes together is empowering. It helps to facilitate belonging, a We-mode as we define those pathways collectively, and embark upon them together. And by motivating action, by providing a reason to act, envisioning futures to move towards also builds efficacy.

Feeling Our Way to Change

This photograph was taken during an art showing of Lori-Ann Claerhout's work, held in Lori's hometown of Athabasca, Alberta, in the Fall of 2023. One of the things I love about Lori's craft is her use of art to express, process, and share her emotional responses to the climate emergency. Here is how Lori describes this work:

The Dreamland Quilt contains birch bark harvested from firewood burned in the stove that heats my house. Previously recorded dreams are typed onto semi-transparent paper and applied, where 'imperfections' like typos, misspellings, and the cutting off of text all point to how dreams can be fractured, ethereal, and difficult to capture. The imperfections come to reflect how thoughts of both despair and hope are not perfect/not complete/not accurate: they are flawed too.

Photo by Kelsey McMillan, reproduced with permission. Check out more of her photography here: https://www.kelsey-mcmillan.com/

This is the hardest one. Engaging in individual or collective action does not involve only our minds, it involves our bodies. In Western cultures, we have been indoctrinated with the belief that to get anything done, to achieve an objective, we must be 'rational,' 'strategic,' we must set our emotions to the side of our desk, or better yet, lock them up in the filing cabinet. How's that working? Great for a military objective, maybe business objectives too, at least in the short term, although there are often enormous personal costs. Soldiers forced to pursue their targets without regard for the emotional impacts of doing so return home with post-traumatic stress disorder;

businesspeople who prioritize the pursuit of profit, perhaps setting aside one's personal moral compass to do so, and without building reciprocal, trusting relationships with customers, carry heavy burdens that they bring home, leading to broken, dysfunctional families.

Pursuing objectives with rationality alone is the last thing we need to confront the climate emergency. Our relations with each other, and with the climate, are quintessentially affective, intimate. Disrupting any of the inaction pathways described in Chapter 6 requires an intervention not at the rational level, not with new education and information campaigns, but at the emotional level. And even those already embracing climate action need to attend to our emotionality, in order for that action to be sustained. First, we need to avoid the strong impulse to run away from fear. The climate emergency is incredibly scary: fear of direct, tangible impacts like extreme weather, and the perhaps less immediately tangible and more slowly moving impacts like economic breakdown. But change itself is scary too: change means uncertainty, it means doing things in new ways, adopting new beliefs; it means letting go.

Second, we need to prepare ourselves for grief, to create space for mourning, in our private spaces and in our community spaces. Western cultures also get a pretty low mark in this realm. Grief, like anxiety, is treated as a problem, just another malady to conquer, because a productive society has no place for grief. And, we are too often left to 'conquer' it alone. Just get over it! We are told, even by well-meaning friends. How do we 'get over it'? That too involves individualized solutions, often provided by the alcohol, entertainment, and pharmaceutical industries. We are taught, in other words, to look away. If we just keep our minds busy enough, those heavy feelings will eventually go away, right? They might well go away, and along with them, our capacity to remember—to honour—our losses, and perhaps also the capacity to honour, and love, the valued beings and elements that remain. If these are the skills we have at our disposal to deal with grief, we are not going to make it. Not only because of the scale of loss in store, but also because, if we lose our capacity to grieve, we lose the capacity to love. Love and grief are the two faces of empathy.

Next, we need to be prepared to move into the negative emotions that can surface when we pursue change. There are negative emotions that can emerge when we build relations with others, when those others are so very different than ourselves, and particularly when previous relations with those others may have been associated with distrust, inequity, and exploitation. We will also need to brace ourselves for when we confront the antagonism and hate expressed by resisters, as anyone who has spoken out about the climate emergency can attest. And we need to expect that we will face failures, perhaps many failures, on our journey towards transformation. Like grief and anxiety, our inclination is to avoid situations that generate negative emotions. So, because pursuing settler-Indigenous reconciliation—one example of a set of relationships that is crucial to addressing the climate emergency—inevitably involves painful recollections, guilt, anger, and distrust, too many of us turn away. We are inclined to only try new things that have a low probability of failure, to avoid the negative feelings, shame in particular, that are likely to emerge. And resistors who have bought into the denial tropes spouted by fossil fuels advocates have been enormously effective in silencing many climate advocates who choose to steer clear of their vitriol. Avoiding these negative emotions, however, amounts to inaction.

How do we then learn to embrace, to sit with, our negative emotions, including grief, and the emotional toll associated with taking action? We avoid the disabling tendencies of negative emotions by sharing them, by sitting with them together, rather than alone. And we do so by cultivating positive emotions, emotions that can energize, and counterbalance the burden of negative emotions. Sharing negative emotions can, in and of itself, be a source of new coalitions and solidarities that offer positive emotional rewards. Imagining futures to move towards can generate positive emotions, most crucially hope. But cultivating positive emotions that can enable us to feel our way to change requires a concerted, proactive approach; it requires that attending to the emotional wellbeing of ourselves and others becomes a central task in our change-making projects, in addition to all the creative and disruptive tactics deployed to promote the curtailment of fossil fuels, and the pursuit of justice. As with the other lofty goals I have set out here, feeling our way to change together facilitates the three basic needs with which we began this chapter: reflexivity—as we sit with our emotions; efficacy—by cultivating the emotional and hence cognitive capacity required for committing oneself to a course of action; and empathy—for others, but even more so, for ourselves.

Works Cited

Adams, Ian, Kristin Hurst, and Nicole D. Sintov. "Experienced Guilt, but Not Pride, Mediates the Effect of Feedback on pro-Environmental Behavior." *Journal of Environmental Psychology* 71 (2020): 101476. https://doi.org/10.1016/j.jenvp.2020.101476.
Adams, Vincanne. "The Other Road to Serfdom: Recovery by the Market and the Affect Economy in New Orleans." *Public Culture* 24, no. 1 (2012): 185–216. https://doi.org/10.1215/08992363-1443601.
Aguilar-Luzón, María Del Carmen, Beatriz Carmona, and Ana Loureiro. "Future Actions towards Climate Change: The Role of Threat Perception and Emotions." *European Journal of Sustainable Development* 12, no. 4 (2023): 379–98. https://doi.org/10.14207/ejsd.2023.v12n4p379.
Ahmed, Sara. "Collective Feelings: Or, the Impressions Left by Others." *Theory, Culture & Society* 21, no. 2 (2004): 25–42. https://doi.org/10.1177/0263276404042133.
Al-Hassani, Ruba A. "Iraq's October Revolution: Between Structures of Patriarchy and Emotion." In *The Palgrave Handbook of Gender, Media and Communication in the Middle East and North Africa*, edited by Loubna H. Skalli and Nahed Eltantawy, 107–25. Cham: Springer International Publishing, 2023. https://doi.org/10.1007/978-3-031-11980-4_7.
Albright, Elizabeth A., and Deserai Crow. "Beliefs about Climate Change in the Aftermath of Extreme Flooding." *Climatic Change* 155, no. 1 (2019): 1–17. https://doi.org/10.1007/s10584-019-02461-2.
Alfred, Taiaiake, and Jeff Corntassel. "Being Indigenous: Resurgences against Contemporary Colonialism." *Government and Opposition* 40, no. 4 (2005): 597–614. https://doi.org/10.1111/j.1477-7053.2005.00166.x.
Alshamsi, Aamena, Fabio Pianesi, Bruno Lepri, Alex Pentland, and Iyad Rahwan. "Beyond Contagion: Reality Mining Reveals Complex Patterns of Social Influence." Edited by Chris T. Bauch. *Plos One* 10, no. 8 (2015): e0135740. https://doi.org/10.1371/journal.pone.0135740.
Amel, Elise, Christie Manning, Britain Scott, and Susan Koger. "Beyond the Roots of Human Inaction: Fostering Collective Effort toward Ecosystem Conservation." *Science* 356, April 21 (2017): 275–79.
Andrews, Jeffrey, and Debra Davidson. "Cell-Gazing Into the Future: What Genes, Homo Heidelbergensis, and Punishment Tell Us About Our Adaptive Capacity." *Sustainability* 5, no. 2 (2013): 560–69. https://doi.org/10.3390/su5020560.
Anfinson, Kellan. "How to Tell the Truth about Climate Change." *Environmental Politics* 27, no. 2 (2018): 209–27. https://doi.org/10.1080/09644016.2017.1413723.

Arcaya, Mariana, Ethan J. Raker, and Mary C. Waters. "The Social Consequences of Disasters: Individual and Community Change." *Annual Review of Sociology* 46 (2020): 671–91.

Arce-García, Sergio, Jesús Díaz-Campo, and Belén Cambronero-Saiz. "Online Hate Speech and Emotions on Twitter: A Case Study of Greta Thunberg at the UN Climate Change Conference COP25 in 2019." *Social Network Analysis and Mining* 13, no. 1 (2023): 48. https://doi.org/10.1007/s13278-023-01052-5.

Archer, Margaret S. *Structure, Agency, and the Internal Conversation.* Cambridge; New York: Cambridge University Press, 2003.

Archer, Margaret Scotford. *Being Human: The Problem of Agency.* Cambridge; New York: Cambridge University Press, 2000.

Archer, Margaret S. *Realist Social Theory: The Morphogenetic Approach.* 1st ed. Cambridge, England: Cambridge University Press, 1995.

Aldrich, Daniel P., and Michelle A. Meyer. "Social Capital and Community Resilience." *American Behavioral Scientist* 59, no. 2 (2015): 254–69. https://doi.org/10.1177/0002764214550299.

Asgarizadeh, Zahra, Robert Gifford, and Lauren Colborne. "Predicting Climate Change Anxiety." *Journal of Environmental Psychology* 90 (2023): 102087. https://doi.org/10.1016/j.jenvp.2023.102087.

Bacon, J. M. "Settler Colonialism as Eco-Social Structure and the Production of Colonial Ecological Violence." *Environmental Sociology* 5, no. 1 (2019): 59–69. https://doi.org/10.1080/23251042.2018.1474725.

Ballew, Matthew T., Seth A. Rosenthal, Matthew H. Goldberg, Abel Gustafson, John E. Kotcher, Edward W. Maibach, and Anthony Leiserowitz. "Beliefs about Others' Global Warming Beliefs: The Role of Party Affiliation and Opinion Deviance." *Journal of Environmental Psychology* 70 (2020): 101466. https://doi.org/10.1016/j.jenvp.2020.101466.

Bandura, Albert. *Self-Efficacy: The Exercise of Control.* New York: W.H. Freeman, 1997.

Bar-Tal, Daniel, Eran Halperin, and Joseph De Rivera. "Collective Emotions in Conflict Situations: Societal Implications." *Journal of Social Issues* 63, no. 2 (2007): 441–60. https://doi.org/10.1111/j.1540-4560.2007.00518.x.

Barbalet, Jack M. *Emotion, Social Theory, and Social Structure.* Cambridge: Cambridge University Press, 1998.

Barlow, David H. *Anxiety and Its Disorders: The Nature and Treatment of Anxiety and Panic.* 2nd ed. Paperback ed. New York, NY: Guilford Press, 2004.

Batson, C. Daniel. "These Things Called Empathy: Eight Related but Distinct Phenomena." In *The Social Neuroscience of Empathy*, edited by Jean Decety and William Ickes, 3–16. The MIT Press, 2009. https://doi.org/10.7551/mitpress/9780262012973.003.0002.

Batson, C. Daniel. "The Empathy-Altruism Hypothesis: Issues and Implications." In *Empathy: From Bench to Bedside*, edited by Jean Decety, 41–54. Cambridge, MA: MIT Press, 2012.

Bauman, Zygmunt. *Liquid Fear.* Reprinted. Cambridge: Polity Press, 2007.

Bauman, Zygmunt. *Liquid Modernity.* Cambridge; Malden, MA: Polity Press; Blackwell, 2000.

Beck, Christopher S. "Attending to the Full Moral Landscape: The Role of Affect in Revealing Obligations to the Other-Than-Human World." *The Arbutus Review* 14, no. 1 (2023). https://doi.org/10.18357/tar141202321365.

Beck, Ulrich. "Living in the World Risk Society: A Hobhouse Memorial Public Lecture given on Wednesday 15 February 2006 at the London School of Economics." *Economy and Society* 35, no. 3 (2006): 329–45. https://doi.org/10.1080/03085140600844902.

Beck, Ulrich. *Risk Society: Towards a New Modernity*. Theory, Culture & Society. London; Newbury Park, CA: Sage Publications, 1992.
Beck, Ulrich, Anders Blok, David Tyfield, and Joy Yueyue Zhang. "Cosmopolitan Communities of Climate Risk: Conceptual and Empirical Suggestions for a New Research Agenda." *Global Networks* 13, no. 1 (2013): 1–21. https://doi.org/10.1111/glob.12001.
Bednar, Jenna. "Polarization, Diversity, and Democratic Robustness." *Proceedings of the National Academy of Sciences* 118, no. 50 (2021): e2113843118. https://doi.org/10.1073/pnas.2113843118.
Benegal, Salil, Flávio Azevedo, and Mirya R. Holman. "Race, Ethnicity, and Support for Climate Policy." *Environmental Research Letters* 17, no. 11 (2022): 114060. https://doi.org/10.1088/1748-9326/aca0ac.
Bericat, Eduardo. "The Sociology of Emotions: Four Decades of Progress." *Current Sociology* 64, no. 3 (2016): 491–513. https://doi.org/10.1177/0011392115588355.
Berry, Helen Louise, Kathryn Bowen, and Tord Kjellstrom. "Climate Change and Mental Health: A Causal Pathways Framework." *International Journal of Public Health* 55, no. 2 (2010): 123–32. https://doi.org/10.1007/s00038-009-0112-0.
Bieniek-Tobasco, Ashley, Sabrina McCormick, Rajiv N. Rimal, Cherise B. Harrington, Madelyn Shafer, and Hina Shaikh. "Communicating Climate Change through Documentary Film: Imagery, Emotion, and Efficacy." *Climatic Change* 154, no. 1–2 (2019): 1–18. https://doi.org/10.1007/s10584-019-02408-7.
Bjork-James, Sophie, and Josef Barla. "A Climate of Misogyny: Gender, Politics of Ignorance, and Climate Change Denial – An Interview with Katharine Hayhoe." *Australian Feminist Studies* 36, no. 110 (2021): 388–95. https://doi.org/10.1080/08164649.2022.2056873.
Black, Simon, Ian Parry, and Nate Vernon. "Fossil Fuel Subsidies Surged to Record $7 Trillion." *IMF Blog (blog)*, 2023.
Blau, Peter M. *Inequality and Heterogeneity: A Primitive Theory of Social Structure*. Free Press, 1977.
Blau, Peter M. *Structural Contexts of Opportunities*. Chicago, IL: University of Chicago Press, 1994.
Blau, Peter M., and Joseph E. Schwartz. *Crosscutting Social Circles*. New York, NY: Academic Press, 1984.
Blenkinsop, Sean, Laura Piersol, and Michael De Danann Sitka-Sage. "Boys Being Boys: Eco-Double Consciousness, Splash Violence, and Environmental Education." *The Journal of Environmental Education* 49, no. 4 (2018): 350–56. https://doi.org/10.1080/00958964.2017.1364213.
Bloodhart, Brittany, Janet K. Swim, and Elaine Dicicco. "'Be Worried, Be VERY Worried:' Preferences for and Impacts of Negative Emotional Climate Change Communication." *Frontiers in Communication* 3 (2019): 63. https://doi.org/10.3389/fcomm.2018.00063.
Bloom, Paul. "Empathy and Its Discontents." *Trends in Cognitive Sciences* 21, no. 1 (2017): 24–31. https://doi.org/10.1016/j.tics.2016.11.004.
Blumenfeld, Jacob. "Climate Barbarism: Adapting to a Wrong World." *Constellations* 30, no. 2 (2023): 162–78. https://doi.org/10.1111/1467-8675.12596.
Boehm, Christopher. *Moral Origins: The Evolution of Virtue, Altruism, and Shame*. New York, NY: Basic Books, 2012.
Bosca, Hannah Della. "Comfort in Chaos: A Sensory Account of Climate Change Denial." *Environment and Planning D: Society and Space* 41, no. 1 (2023): 170–87. https://doi.org/10.1177/02637758231153399.
Bostrom, Ann, Adam L. Hayes, and Katherine M. Crosman. "Efficacy, Action, and Support for Reducing Climate Change Risks." *Risk Analysis* 39, no. 4 (2019): 805–28. https://doi.org/10.1111/risa.13210.

Boudet, Hilary, Leanne Giordono, Chad Zanocco, Hannah Satein, and Hannah Whitley. "Event Attribution and Partisanship Shape Local Discussion of Climate Change after Extreme Weather." *Nature Climate Change* 10, no. 1 (2020): 69–76. https://doi.org/10.1038/s41558-019-0641-3.

Bouman, Thijs, Mark Verschoor, Casper J. Albers, Gisela Böhm, Stephen D. Fisher, Wouter Poortinga, Lorraine Whitmarsh, and Linda Steg. "When Worry about Climate Change Leads to Climate Action: How Values, Worry and Personal Responsibility Relate to Various Climate Actions." *Global Environmental Change* 62 (2020): 102061. https://doi.org/10.1016/j.gloenvcha.2020.102061.

Bourdieu, Pierre. *Outline of a Theory of Practice*. Cambridge: Cambridge University Press, 2019.

Boyd, Robert, and Peter J. Richerson. *Culture and the Evolutionary Process*. Chicago, IL: University of Chicago Press, 1985.

Bradley, Margaret M., Bruce N. Cuthbert, and Peter J. Lang. "Picture Media and Emotion: Effects of a Sustained Affective Context." *Psychophysiology* 33, no. 6 (1996): 662–70. https://doi.org/10.1111/j.1469-8986.1996.tb02362.x.

Bransford, John, National Research Council (U.S.), and National Research Council (U.S.), eds. *How People Learn: Brain, Mind, Experience, and School*. Expanded ed. Washington, DC: National Academy Press, 2000.

Brave Heart, Maria Yellow Horse, Jennifer Elkins, Greg Tafoya, Doreen Bird, and Melina Salvador. "Wicasa Was'aka: Restoring the Traditional Strength of American Indian Boys and Men." *American Journal of Public Health* 102, no. S2 (2012): S177–83. https://doi.org/10.2105/AJPH.2011.300511.

Breggin, Peter R. "The Biological Evolution of Guilt, Shame and Anxiety: A New Theory of Negative Legacy Emotions." *Medical Hypotheses* 85, no. 1 (2015): 17–24. https://doi.org/10.1016/j.mehy.2015.03.015.

Brescoll, Victoria L., and Eric Luis Uhlmann. "Can an Angry Woman Get Ahead?: Status Conferral, Gender, and Expression of Emotion in the Workplace." *Psychological Science* 19, no. 3 (2008): 268–75. https://doi.org/10.1111/j.1467-9280.2008.02079.x.

Breunig, Mary, and Constance Russell. "Long-Term Impacts of Two Secondary School Environmental Studies Programs on Environmental Behaviour: The Shadows of Patriarchy and Neoliberalism." *Environmental Education Research* 26, no. 5 (2020): 701–15. https://doi.org/10.1080/13504622.2020.1749236.

Brick, Cameron, Anna Bosshard, and Lorraine Whitmarsh. "Motivation and Climate Change: A Review." *Current Opinion in Psychology* 42 (2021): 82–88. https://doi.org/10.1016/j.copsyc.2021.04.001.

Bright, Maria L., and Chris Eames. "From Apathy through Anxiety to Action: Emotions as Motivators for Youth Climate Strike Leaders." *Australian Journal of Environmental Education* 38, no. 1 (2022): 13–25. https://doi.org/10.1017/aee.2021.22.

Brown, Katrina, W. Neil Adger, Patrick Devine-Wright, John M. Anderies, Stewart Barr, Francois Bousquet, Catherine Butler, Louisa Evans, Nadine Marshall, and Tara Quinn. "Empathy, Place and Identity Interactions for Sustainability." *Global Environmental Change* 56 (2019): 11–17. https://doi.org/10.1016/j.gloenvcha.2019.03.003.

Brulle, Robert J., and Kari Marie Norgaard. "Avoiding Cultural Trauma: Climate Change and Social Inertia." *Environmental Politics* 28, no. 5 (2019): 886–908. https://doi.org/10.1080/09644016.2018.1562138.

Bury, Simon M., Michael Wenzel, and Lydia Woodyatt. "Against the Odds: Hope as an Antecedent of Support for Climate Change Action." *British Journal of Social Psychology* 59, no. 2 (2020): 289–310. https://doi.org/10.1111/bjso.12343.

Bury, Simon M., Michael Wenzel, and Lydia Woodyatt. "Confusing Hope and Optimism When Prospects Are Good: A Matter of Language Pragmatics or Conceptual Equivalence?" *Motivation and Emotion* 43, no. 3 (2019): 483–92. https://doi.org/10.1007/s11031-018-9746-7.
Carver, Charles S., Michael F. Scheier, and Suzanne C. Segerstrom. "Optimism." *Clinical Psychology Review* 30, no. 7 (2010): 879–89. https://doi.org/10.1016/j.cpr.2010.01.006.
Cassegård, Carl, and Håkan Thörn. "Toward a Postapocalyptic Environmentalism? Responses to Loss and Visions of the Future in Climate Activism." *Environment and Planning E: Nature and Space* 1, no. 4 (2018): 561–78. https://doi.org/10.1177/2514848618793331.
Chandio, Abbas Ali, Yuansheng Jiang, Asad Amin, Munir Ahmad, Waqar Akram, and Fayyaz Ahmad. "Climate Change and Food Security of South Asia: Fresh Evidence from a Policy Perspective Using Novel Empirical Analysis." *Journal of Environmental Planning and Management* 66, no. 1 (2023): 169–90. https://doi.org/10.1080/09640568.2021.1980378.
Chapman, Daniel A., Adam Corner, Robin Webster, and Ezra M. Markowitz. "Climate Visuals: A Mixed Methods Investigation of Public Perceptions of Climate Images in Three Countries." *Global Environmental Change* 41 (2016): 172–82. https://doi.org/10.1016/j.gloenvcha.2016.10.003.
Chen, Mei-Fang. "Self-Efficacy or Collective Efficacy within the Cognitive Theory of Stress Model: Which More Effectively Explains People's Self-Reported Proenvironmental Behavior?" *Journal of Environmental Psychology* 42 (2015): 66–75. https://doi.org/10.1016/j.jenvp.2015.02.002.
Chen, Shuquan, Rohini Bagrodia, Charlotte C. Pfeffer, Laura Meli, and George A. Bonanno. "Anxiety and Resilience in the Face of Natural Disasters Associated with Climate Change: A Review and Methodological Critique." *Journal of Anxiety Disorders* 76 (2020): 102297. https://doi.org/10.1016/j.janxdis.2020.102297.
Christakis, Nicholas A. *Blueprint: The Evolutionary Origins of a Good Society*. New York, NY: Little Brown Spark, 2019.
Christophe, Véronique, and Bernard Rimé. "Exposure to the Social Sharing of Emotion: Emotional Impact, Listener Responses and the Secondary Social Sharing." *European Journal of Social Psychology* 27 (1997): 37–54.
Cikara, Mina, and Jay J. Van Bavel. "The Neuroscience of Intergroup Relations: An Integrative Review." *Perspectives on Psychological Science* 9, no. 3 (2014): 245–74. https://doi.org/10.1177/1745691614527464.
Clark, Candace. *Misery and Company: Sympathy in Everyday Life*. Chicago, IL: University of Chicago Press, 1997.
Clark, Kylie, and Aubrie Adams. "Altering Attitudes on Climate Change: Testing the Effect of Time Orientation and Motivation Framing." *CSU Journal of Sustainability and Climate Change* 3, no. 1 (2023). https://doi.org/10.55671/2771-5582.1019.
Clayton, Susan, and Bryan T. Karazsia. "Development and Validation of a Measure of Climate Change Anxiety." *Journal of Environmental Psychology* 69 (2020): 101434. https://doi.org/10.1016/j.jenvp.2020.101434.
Clot-Garrell, Anna. "Voices of Emergency: Imagined Climate Futures and Forms of Collective Action." *Current Sociology* (2023): 00113921231182179. https://doi.org/10.1177/00113921231182179.
Coffey, Yumiko, Navjot Bhullar, Joanne Durkin, Md Shahidul Islam, and Kim Usher. "Understanding Eco-Anxiety: A Systematic Scoping Review of Current Literature and Identified Knowledge Gaps." *The Journal of Climate Change and Health* 3 (2021): 100047. https://doi.org/10.1016/j.joclim.2021.100047.

Works Cited

Cohen-Chen, Smadar, and Martijn Van Zomeren. "Yes We Can? Group Efficacy Beliefs Predict Collective Action, but Only When Hope Is High." *Journal of Experimental Social Psychology* 77 (2018): 50–59. https://doi.org/10.1016/j.jesp.2018.03.016.

Colleoni, Elanor, Alessandro Rozza, and Adam Arvidsson. "Echo Chamber or Public Sphere? Predicting Political Orientation and Measuring Political Homophily in Twitter Using Big Data: Political Homophily on Twitter." *Journal of Communication* 64, no. 2 (2014): 317–32. https://doi.org/10.1111/jcom.12084.

Collins, Patricia H. "Black Feminist Epistemology." In *Contemporary Sociological Theory*, edited by Jonathan H. Turner, 3rd ed., 407–17. Chichester: Wiley-Blackwell, 2012.

Collins, Randall. *Interaction Ritual Chains*. Princeton, NJ: Princeton University Press, 2004.

Conte, Beatrice, Tobias Brosch, and Ulf J. J. Hahnel. "Initial Evidence for a Systematic Link between Core Values and Emotional Experiences in Environmental Situations." *Journal of Environmental Psychology* 88 (2023): 102026. https://doi.org/10.1016/j.jenvp.2023.102026.

Cormick, Craig. "Who Doesn't Love a Good Story? — What Neuroscience Tells about How We Respond to Narratives." *Journal of Science Communication* 18, no. 5 (2019): Y01. https://doi.org/10.22323/2.18050401.

Coviello, Lorenzo, Yunkyu Sohn, Adam D. I. Kramer, Cameron Marlow, Massimo Franceschetti, Nicholas A. Christakis, and James H. Fowler. "Detecting Emotional Contagion in Massive Social Networks." *Plos One* 9, no. 3 (2014): e90315.

Cox, Daniel T. C., Danielle F. Shanahan, Hannah L. Hudson, Kate E. Plummer, Gavin M. Siriwardena, Richard A. Fuller, Karen Anderson, Steven Hancock, and Kevin J. Gaston. "Doses of Neighborhood Nature: The Benefits for Mental Health of Living with Nature." *BioScience* 67, no. 2 (2017): 147–55. https://doi.org/10.1093/biosci/biw173.

Cozolino, Louis. *The Neuroscience of Human Relationships: Attachment and the Developing Social Brain*. New York, NY: W. W. Norton, 2006.

Creed, Douglas W. E., Bryant Ashley Hudson, Gerardo A. Okhuysen, and Kristin Smith-Crowe. "Swimming in a Sea of Shame: Incorporating Emotion into Explanations of Institutional Reproduction and Change." *Academy of Management Review* 39, no. 3 (2014): 275–301. https://doi.org/10.5465/amr.2012.0074.

Cruz, Shannon M., and Andrew C. High. "Psychometric Properties of the Climate Change Anxiety Scale." *Journal of Environmental Psychology* 84 (2022): 101905. https://doi.org/10.1016/j.jenvp.2022.101905.

Cuartas, Jorge, Amiya Bhatia, Daniel Carter, Lucie Cluver, Carolina Coll, Elizabeth Donger, Catherine E. Draper, et al. "Climate Change Is a Threat Multiplier for Violence against Children." *Child Abuse & Neglect* (2023): 106430. https://doi.org/10.1016/j.chiabu.2023.106430.

Da Costa, Dia. "Cruel Pessimism and Waiting for Belonging: Towards a Global Political Economy of Affect." *Cultural Studies* 30, no. 1 (2016): 1–23. https://doi.org/10.1080/09502386.2014.899607.

Daggett, Cara New. *The Birth of Energy: Fossil Fuels, Thermodynamics, and the Politics of Work*. Elements. Durham: Duke University Press, 2019.

Damásio, Antonio. *Descartes' Error: Emotion, Reason and the Human Brain*. New York, NY: G.P. Putman, 1994.

Darwin, Charles. *The Origin of Species*. New York, NY: Gramercy Books, 1979.

Davidson, Debra J. "Analysing Responses to Climate Change through the Lens of Reflexivity." *The British Journal of Sociology* 63, no. 4 (2012): 616–40. https://doi.org/10.1111/j.1468-4446.2012.01429.x.

Davidson, Debra J., Anthony Fisher, and Gwendolyn Blue. "Missed Opportunities: The Absence of Climate Change in Media Coverage of Forest Fire Events in Alberta." *Climatic Change* 153, no. 1–2 (2019): 165–79. https://doi.org/10.1007/s10584-019-02378-w.

Davidson, Debra J., Curtis Rollins, Lianne Lefsrud, Sven Anders, and Andreas Hamann. "Just Don't Call It Climate Change: Climate-Skeptic Farmer Adoption of Climate-Mitigative Practices." *Environmental Research Letters* 14, no. 3 (2019): 034015. https://doi.org/10.1088/1748-9326/aafa30.

Davis, Mark H., Carol Luce, and Stephen J. Kraus. "The Heritability of Characteristics Associated with Dispositional Empathy." *Journal of Personality* 62, no. 3 (1994): 369–91.

De Silva, M. M. G. T., and Akiyuki Kawasaki. "Socioeconomic Vulnerability to Disaster Risk: A Case Study of Flood and Drought Impact in a Rural Sri Lankan Community." *Ecological Economics* 152 (2018): 131–40. https://doi.org/10.1016/j.ecolecon.2018.05.010.

De Waal, Frans B. M. "Putting the Altruism Back into Altruism: The Evolution of Empathy." *Annual Review of Psychology* 59, no. 1 (2008): 279–300. https://doi.org/10.1146/annurev.psych.59.103006.093625.

Dean, Angela J., and Kerrie A. Wilson. "Relationships between Hope, Optimism, and Conservation Engagement." *Conservation Biology* 37, no. 2 (2023): e14009. https://doi.org/10.1111/cobi.14020.

Dean, Rebecca M. "Social Change and Hunting during the Pueblo III to Pueblo IV Transition, East-Central Arizona." *Journal of Field Archaeology* 28, no. 3/4 (2001): 271–85.

Decety, Jean. "Dissecting the Neural Mechanisms Mediating Empathy." *Emotion Review* 3, no. 1 (2011): 92–108. https://doi.org/10.1177/1754073910374662.

Decety, Jean, and Jason M. Cowell. "Empathy, Justice, and Moral Behavior." *AJOB Neuroscience* 6, no. 3 (2015): 3–14. https://doi.org/10.1080/21507740.2015.1047055.

Decety, Jean, and Philip L. Jackson. "The Functional Architecture of Human Empathy." *Behavioral and Cognitive Neuroscience Reviews* 3, no. 2 (2004): 71–100. https://doi.org/10.1177/1534582304267187.

Decety, Jean, and Meghan Meyer. "From Emotion Resonance to Empathic Understanding: A Social Developmental Neuroscience Account." *Development and Psychopathology* 20, no. 4 (2008): 1053–80. https://doi.org/10.1017/S0954579408000503.

Dehingia, Nabamallika, Lotus McDougal, Jay G. Silverman, Elizabeth Reed, Lianne Urada, Julian McAuley, Abhishek Singh, and Anita Raj. "Climate and Gender: Association between Droughts and Intimate Partner Violence in India." *American Journal of Epidemiology* (2023): kwad222. https://doi.org/10.1093/aje/kwad222.

Denny, Riva C. H., Julia Marchese, and A. Paige Fischer. "Severe Weather Experience and Climate Change Belief among Small Woodland Owners: A Study of Reciprocal Effects." *Weather, Climate, and Society* 14, no. 4 (2022): 1065–82. https://doi.org/10.1175/WCAS-D-21-0176.1.

Denworth, Lydia. "'Hyperscans' Show How Brains Sync as People Interact." *Scientific American April* 10 (2019).

Dessí, Roberta, and Benoît Monin. "'Noblesse Oblige? Moral Identity and Prosocial Behavior in the Face of Selfishness.'" Toulouse School of Economics, 2012. Working Paper Series 12-347.

Dhont, Kristof, Gordon Hodson, Kimberly Costello, and Cara C. MacInnis. "Social Dominance Orientation Connects Prejudicial Human–Human and Human–Animal Relations." *Personality and Individual Differences* 61–62 (2014): 105–8. https://doi.org/10.1016/j.paid.2013.12.020.

Diamond, Jared M. *Collapse: How Societies Choose to Fail or Succeed.* Harmondsworth: Penguin Books, 2006.
Diamond, Emily, and Kaitlin Urbanski. "The Impact of Message Valence on Climate Change Attitudes: A Longitudinal Experiment." *Environmental Communication* 16, no. 8 (2022): 1046–58. https://doi.org/10.1080/17524032.2022.2151486.
Ditlevsen, Peter, and Susanne Ditlevsen. "Warning of a Forthcoming Collapse of the Atlantic Meridional Overturning Circulation." *Nature Communications* 14, no. 1 (2023): 4254. https://doi.org/10.1038/s41467-023-39810-w.
Doherty, Fiona C., Smitha Rao, and Angelise R. Radney. "Association between Child, Early, and Forced Marriage and Extreme Weather Events: A Mixed-Methods Systematic Review." *International Social Work* (2023): 00208728231186006. https://doi.org/10.1177/00208728231186006.
Downs, Julie S. "Prescriptive Scientific Narratives for Communicating Usable Science." *Proceedings of the National Academy of Sciences* 111, no. Supplement 4 (2014): 13627–33.
Duckitt, John, and Chris G. Sibley. "Personality, Ideology, Prejudice, and Politics: A Dual-Process Motivational Model." *Journal of Personality* 78, no. 6 (2010): 1861–94. https://doi.org/10.1111/j.1467-6494.2010.00672.x.
Durkheim, Emile. *The Elementary Forms of the Religious Life.* London: Allen and Unwin, 1976.
Eisenberg, Nancy, Tracy L. Spinrad, and Adrienne Morris. "Empathy-Related Responding in Children." In *Handbook of Moral Development*, edited by Melanie Killen and Judith G. Smetana, 184–207. Psychology Press, 2006.
Erikson, Kai T. *Everything in Its Path: Destruction of Community in the Buffalo Creek Flood.* Nachdr. New York: Simon & Schuster, 1978.
Farmer, Harry, and Lara Maister. "Putting Ourselves in Another's Skin: Using the Plasticity of Self-Perception to Enhance Empathy and Decrease Prejudice." *Social Justice Research* 30 (2017): 323–54.
Ferguson, Mark A., and Nyla R. Branscombe. "Collective Guilt Mediates the Effect of Beliefs about Global Warming on Willingness to Engage in Mitigation Behavior." *Journal of Environmental Psychology* 30, no. 2 (2010): 135–42. https://doi.org/10.1016/j.jenvp.2009.11.010.
Ferguson, Susan, Genevieve LeBaron, Angela Dimitrakaki, and Sara R. Farris. "Introduction." *Historical Materialism* 24, no. 2 (2016): 25–37. https://doi.org/10.1163/1569206X-12341469.
Ferran, Íngrid V. "Empathy, Emotional Sharing and Feelings in Stein's Early Work." *Human Studies* 38, no. 4 (2015): 481–502. https://doi.org/10.1007/s10746-015-9346-4.
Ferrarese, Estelle. "Precarity of Work, Precarity of Moral Dispositions: Concern for Others in the Era of 'Emotional' Capitalism." *Women's Studies Quarterly* 45, no. 3/4 (2017): 178–92.
Findlater, K. M., S. D. Donner, T. Satterfield, and M. Kandlikar. "Integration Anxiety: The Cognitive Isolation of Climate Change." *Global Environmental Change* 50 (2018): 178–89. https://doi.org/10.1016/j.gloenvcha.2018.02.010.
Fineman, Stephen, and Andrew Sturdy. "The Emotions of Control: A Qualitative Exploration of Environmental Regulation." *Human Relations* 52, no. 5 (1999): 631–63.
Fisher, Walter R. *Human Communication as Narration: Toward a Philosophy of Reason, Value, and Action.* University of South Carolina Press, 2021. https://doi.org/10.2307/j.ctv1nwbqtk.
Flam, Helena. "Emotional 'Man': I. The Emotional 'man' and the Problem of Collective Action." *International Sociology* 5, no. 1 (1990): 39–56.

Flowers, Rachel. "Refusal to Forgive: Indigenous Women's Love and Rage." *Decolonization: Indigeneity, Education & Society* 4, no. 2 (2015): 32–49.

Flynn, James, Paul Slovic, and C. K. Mertz. "Gender, Race, and Perception of Environmental Health Risks." *Risk Analysis* 14, no. 6 (1994): 1101–8. https://doi.org/10.1111/j.1539-6924.1994.tb00082.x.

Folbre, Nancy, and Julie A Nelson. "For Love or Money—Or Both?" *Journal of Economic Perspectives* 14, no. 4 (2000): 123–40. https://doi.org/10.1257/jep.14.4.123.

Fowler, James H, and Nicholas A Christakis. "Dynamic Spread of Happiness in a Large Social Network: Longitudinal Analysis over 20 Years in the Framingham Heart Study." *British Medical Journal* 337 (2008): 1–9.

Fox, Nick J. "Emotions, Affects and the Production of Social Life." *The British Journal of Sociology* 66, no. 2 (2015): 301–18. https://doi.org/10.1111/1468-4446.12119.

Fransen, Taryn, Jonas Meckling, Anna Stünzi, Tobias S. Schmidt, Florian Egli, Nicolas Schmid, and Christopher Beaton. "Taking Stock of the Implementation Gap in Climate Policy." *Nature Climate Change* 13, no. 8 (2023): 752–55. https://doi.org/10.1038/s41558-023-01755-9.

Freire, Paulo. *Pedagogy of Hope: Reliving Pedagogy of the Oppressed*. London; New York: Bloomsbury, 2014.

Frijda, Nico H. "The Evolutionary Emergence of What We Call 'Emotions.'" *Cognition and Emotion* 30, no. 4 (2016): 609–20. https://doi.org/10.1080/02699931.2016.1145106.

Fritsche, Immo, Markus Barth, Philipp Jugert, Torsten Masson, and Gerhard Reese. "A Social Identity Model of Pro-Environmental Action (SIMPEA)." *Psychological Review* 125, no. 2 (2018): 245–69. https://doi.org/10.1037/rev0000090.

Fritze, Jessica G., Grant A. Blashki, Susie Burke, and John Wiseman. "Hope, Despair and Transformation: Climate Change and the Promotion of Mental Health and Wellbeing." *International Journal of Mental Health Systems* 2, no. 1 (2008): 13. https://doi.org/10.1186/1752-4458-2-13.

Galway, Lindsay P., Thomas Beery, Chris Buse, and Maya K. Gislason. "What Drives Climate Action in Canada's Provincial North? Exploring the Role of Connectedness to Nature, Climate Worry, and Talking with Friends and Family." *Climate* 9, no. 10 (2021): 146. https://doi.org/10.3390/cli9100146.

Galway, Lindsay P., and Ellen Field. "Climate Emotions and Anxiety among Young People in Canada: A National Survey and Call to Action." *The Journal of Climate Change and Health* 9 (2023): 100204. https://doi.org/10.1016/j.joclim.2023.100204.

Gebhardt, Nadja, Katharina Van Bronswijk, Maxie Bunz, Tobias Müller, Pia Niessen, and Christoph Nikendei. "Scoping Review of Climate Change and Mental Health in Germany – Direct and Indirect Impacts, Vulnerable Groups, Resilience Factors," 2023. https://doi.org/10.25646/11656.

Geiger, Nathaniel, Caitlin R. Bowman, Tracy L. Clouthier, Anthony J. Nelson, and Reginald B. Adams. "Observing Environmental Destruction Stimulates Neural Activation in Networks Associated with Empathic Responses." *Social Justice Research* 30, no. 4 (2017): 300–322. https://doi.org/10.1007/s11211-017-0298-x.

Geiger, Nathaniel, Janet K. Swim, Karen Gasper, John Fraser, and Kate Flinner. "How Do I Feel When I Think about Taking Action? Hope and Boredom, Not Anxiety and Helplessness, Predict Intentions to Take Climate Action." *Journal of Environmental Psychology* 76 (2021): 101649. https://doi.org/10.1016/j.jenvp.2021.101649.

Ghosh, Amitav. *The Nutmeg's Curse: Parables for a Planet in Crisis*. Paperback edition. Chicago: The University of Chicago Press, 2022.

Giddens, Anthony. *The Consequences of Modernity*. 6th pr. Stanford, CA: Stanford Univ. Press, 1997.

Giddens, Anthony. "The Politics of Climate Change." *Policy & Politics* 43, no. 2 (2015): 155–62. https://doi.org/10.1332/030557315X14290856538163.

Gignoux, Jérémie, and Marta Menéndez. "Benefit in the Wake of Disaster: Long-Run Effects of Earthquakes on Welfare in Rural Indonesia." *Journal of Development Economics* 118 (2016): 26–44. https://doi.org/10.1016/j.jdeveco.2015.08.004.

Gilligan, Carol. *In a Different Voice: Psychological Theory and Women's Development*. Cambridge, MA: Harvard University Press, 1982.

Gilligan, Carol, and Naomi Snider. *Why Does Patriarchy Persist?* Cambridge: Polity Press, 2018.

Goffman, Erving. *The Presentation of Self in Everyday Life*. Doubleday, 1959.

Gössling, Stefan, Andreas Humpe, and Thomas Bausch. "Does 'Flight Shame' Affect Social Norms? Changing Perspectives on the Desirability of Air Travel in Germany." *Journal of Cleaner Production* 266 (2020): 122015. https://doi.org/10.1016/j.jclepro.2020.122015.

Gotham, Kevin Fox. "Reinforcing Inequalities: The Impact of the CDBG Program on Post-Katrina Rebuilding." *Housing Policy Debate* 24, no. 1 (2014): 192–212. https://doi.org/10.1080/10511482.2013.840666.

Gotham, Kevin Fox. "Re-Anchoring Capital in Disaster-Devastated Spaces: Financialisation and the Gulf Opportunity (GO) Zone Programme." *Urban Studies* 53, no. 7 (2016): 1362–83. https://doi.org/10.1177/0042098014548117.

Gotham, Kevin Fox, and Miriam Greenberg. *Crisis Cities: Disaster and Redevelopment in New York and New Orleans*. Oxford: Oxford University Press, 2014.

Gould, Stephen J. *An Urchin in the Storm*. New York, NY: W. W. Norton, 1987.

Graeber, David, and D. Wengrow. *The Dawn of Everything: A New History of Humanity*. Toronto, Ontario: Signal, an imprint of McClelland & Stewart, 2021.

Greenaway, Katharine H., Aleksandra Cichocka, Ruth Van Veelen, Tiina Likki, and Nyla R. Branscombe. "Feeling Hopeful Inspires Support for Social Change." *Political Psychology* 37, no. 1 (2016): 89–107. https://doi.org/10.1111/pops.12225.

Gregersen, Thea, Rouven Doran, Gisela Böhm, and Wouter Poortinga. "Outcome Expectancies Moderate the Association between Worry about Climate Change and Personal Energy-Saving Behaviors." Edited by Arkadiusz Piwowar. *PLOS ONE* 16, no. 5 (2021): e0252105. https://doi.org/10.1371/journal.pone.0252105.

Gunderson, Lance H., and Crawford S. Holling, eds. *Panarchy: Understanding Transformations in Human and Natural Systems*. Washington, DC: Island Press, 2002.

Gurven, Michael D. "Broadening Horizons: Sample Diversity and Socioecological Theory Are Essential to the Future of Psychological Science." *Proceedings of the National Academy of Sciences* 115, no. 45 (2018): 11420–27. https://doi.org/10.1073/pnas.1720433115.

Guthridge, Michaela, Paul H. Mason, Tania Penovic, and Melita J. Giummarra. "A Critical Review of Interdisciplinary Perspectives on the Paradox of Prosocial Compared to Antisocial Manifestations of Empathy." *Social Science Information* 59, no. 4 (2020): 632–53. https://doi.org/10.1177/0539018420976946.

Gutsell, Jennifer N., and Michael Inzlicht. "A Neuroaffective Perspective on Why People Fail to Live a Sustainable Lifestyle." In *Encouraging Sustainable Behavior*, edited by H. van Trijp, 137–51. Psychology Press, 2014.

Habermas, Jurgen. *Toward a Rational Society*. Boston: Beacon Press, 1970.

Hahnel, Ulf J. J., Christian Mumenthaler, and Tobias Brosch. "Emotional Foundations of the Public Climate Change Divide." *Climatic Change* 161, no. 1 (2020): 9–19. https://doi.org/10.1007/s10584-019-02552-0.

Hall, Nina, and Lucy Crosby. "Climate Change Impacts on Health in Remote Indigenous Communities in Australia." *International Journal of Environmental Health Research* 32, no. 3 (2022): 487–502. https://doi.org/10.1080/09603123.2020.1777948.

Haltinner, Kristin, Jennifer Ladino, and Dilshani Sarathchandra. "Feeling Skeptical: Worry, Dread, and Support for Environmental Policy among Climate Change Skeptics." *Emotion, Space and Society* 39 (2021): 100790. https://doi.org/10.1016/j.emospa.2021.100790.

Haltinner, Kristin, and Dilshani Sarathchandra. "Climate Change Skepticism as a Psychological Coping Strategy." *Sociology Compass* 12, no. 6 (2018): e12586. https://doi.org/10.1111/soc4.12586.

Hamilton, Lawrence C., Cameron P. Wake, Joel Hartter, Thomas G. Safford, and Alli J. Puchlopek. "Flood Realities, Perceptions and the Depth of Divisions on Climate." *Sociology* 50, no. 5 (2016): 913–33. https://doi.org/10.1177/0038038516648547.

Hansen, James E., Makiko Sato, Leon Simons, Larissa S. Nazarenko, Isabelle Sangha, Pushker Kharecha, James C Zachos, et al. "Global Warming in the Pipeline." *Oxford Open Climate Change* 3, no. 1 (2023): kgad008. https://doi.org/10.1093/oxfclm/kgad008.

Hansla, André, Amelie Gamble, Asgeir Juliusson, and Tommy Gärling. "The Relationships between Awareness of Consequences, Environmental Concern, and Value Orientations." *Journal of Environmental Psychology* 28, no. 1 (2008): 1–9. https://doi.org/10.1016/j.jenvp.2007.08.004.

Harré, Rom. "Social Reality and the Myth of Social Structure." *European Journal of Social Theory* 5, no. 1 (2002): 111–23.

Harth, Nicole S. "Affect, (Group-Based) Emotions, and Climate Change Action." *Current Opinion in Psychology* 42 (2021): 140–44. https://doi.org/10.1016/j.copsyc.2021.07.018.

Hassol, Susan J. "The Right Words Are Crucial to Solving Climate Change – Scientific American." *Scientific American*, February (2023). https://www.scientificamerican.com/article/the-right-words-are-crucial-to-solving-climate-change/.

He, Haozhe, Ryan J. Kramer, Brian J. Soden, and Nadir Jeevanjee. "State Dependence of CO_2 Forcing and Its Implications for Climate Sensitivity." *Science* 382, no. 6674 (2023): 1051–56. https://doi.org/10.1126/science.abq6872.

He, Saike, Xiaolong Zheng, Daniel Zeng, Chuan Luo, and Zhu Zhang. "Exploring Entrainment Patterns of Human Emotion in Social Media." *PLOS ONE* 11, no. 3 (2016): e0150630.

Heath, Chip, Chris Bell, and Emily Sternberg. "Emotional Selection in Memes: The Case of Urban Legends." *Journal of Personality and Social Psychology* 81 (2001): 1028–41.

Helm, Sabrina V., Xiaomin Li, Melissa A. Curran, and Melissa A. Barnett. "Coping Profiles in the Context of Global Environmental Threats: A Person-Centered Approach." *Anxiety, Stress, & Coping* 35, no. 5 (2022): 609–22. https://doi.org/10.1080/10615806.2021.2004132.

Hepp, Johanna, Sina A. Klein, Luisa K. Horsten, Jana Urbild, and Sean P. Lane. "Introduction and Behavioral Validation of the Climate Change Distress and Impairment Scale." *Scientific Reports* 13, no. 1 (2023): 11272. https://doi.org/10.1038/s41598-023-37573-4.

Hickman, Caroline. "We Need to (Find a Way to) Talk about … Eco-Anxiety." *Journal of Social Work Practice* 34, no. 4 (2020): 411–24. https://doi.org/10.1080/02650533.2020.1844166.

Hill, Kim R., Robert S. Walker, Miran Božičević, James Eder, Thomas Headland, Barry Hewlett, A. Magdalena Hurtado, Frank Marlowe, Polly Wiessner, and Brian Wood. "Co-Residence Patterns in Hunter-Gatherer Societies Show Unique

Works Cited

Human Social Structure." *Science* 331, no. 6022 (2011): 1286–89. https://doi.org/10.1126/science.1199071.

Hobson, R. Peter. "On Sharing Experiences." *Development and Psychopathology* 1 (1989): 197–203.

Hochschild, Arlie Russell. *So How's the Family? And Other Essays*. University of California Press, 2013.

Hochschild, Arlie Russell. *The Managed Heart: Commercialization of Human Feeling*. Berkeley, CA: University of California Press, 1983.

Hoffman, Martin L. "Empathy and Prosocial Behavior." In *Handbook of Emotions*, edited by Michael Lewis, Jeannette M. Haviland-Jones, and Lisa F. Barrett, 3rd ed., 440–55. Guilford Press, 2008.

Holleman, Hannah. "De-Naturalizing Ecological Disaster: Colonialism, Racism and the Global Dust Bowl of the 1930s." *The Journal of Peasant Studies* 44, no. 1 (2017): 234–60. https://doi.org/10.1080/03066150.2016.1195375.

Hommel, Bernhard, Jochen Müsseler, Gisa Aschersleben, and Wolfgang Prinz. "The Theory of Event Coding (TEC): A Framework for Perception and Action Planning." *Behavioral and Brain Sciences* 24, no. 5 (2001): 849–78. https://doi.org/10.1017/S0140525X01000103.

Hornsey, Matthew J., Cassandra M. Chapman, and Jacquelyn E. Humphrey. "Climate Skepticism Decreases When the Planet Gets Hotter and Conservative Support Wanes." *Global Environmental Change* 74 (2022): 102492. https://doi.org/10.1016/j.gloenvcha.2022.102492.

Howe, Peter D., Jennifer R. Marlon, Matto Mildenberger, and Brittany S. Shield. "How Will Climate Change Shape Climate Opinion?" *Environmental Research Letters* 14, no. 11 (2019): 113001. https://doi.org/10.1088/1748-9326/ab466a.

Hsiang, Solomon M., Marshall Burke, and Edward Miguel. "Quantifying the Influence of Climate on Human Conflict." *Science* 341, no. 6151 (2013): 1235367. https://doi.org/10.1126/science.1235367.

Hultman, Martin, and Paul M. Pulé. *Ecological Masculinities: Theoretical Foundations and Practical Guidance*. Routledge Studies in Gender and Environments. Abingdon, New York, NY: Routledge, 2018.

Humphrey, Nicholas K. "The Uses of Consciousness." In *Speculations: The Reality Club*, edited by John Brockman, 67–84. New York, NY: Prentice Hall, 1990.

Hurst, Kristin F., and Nicole D. Sintov. "Guilt Consistently Motivates Pro-Environmental Outcomes While Pride Depends on Context." *Journal of Environmental Psychology* 80 (2022): 101776. https://doi.org/10.1016/j.jenvp.2022.101776.

Iacoboni, Marco. *Mirroring People: The New Science of How We Connect with Others*. New York, NY: Farrar, Straus and Giroux, 2008.

Immordino-Yang, Mary Helen, and Antonio Damasio. "We Feel, Therefore We Learn: The Relevance of Affective and Social Neuroscience to Education," *Mind, Brain, and Education* 1, no. 1 (2007): 3–10.

Institute for Economics and Peace. "Over One Billion People at Threat of Being Displaced by 2050 Due to Environmental Change, Conflict and Civil Unrest." Institute for Economics and Peace, September 9, 2020. https://www.prnewswire.com/news-releases/iep-over-one-billion-people-at-threat-of-being-displaced-by-2050-due-to-environmental-change-conflict-and-civil-unrest-301125350.html#:~:text=If%20natural%20disasters%20occur%20at,over%20581%2C000%20recorded%20since%201990.

Intergovernmental Panel On Climate Change (Ipcc). *Climate Change 2022 – Impacts, Adaptation and Vulnerability: Working Group II Contribution to the Sixth Assessment Report of the Intergovernmental Panel on Climate Change*. 1st ed. Cambridge University Press, 2023. https://doi.org/10.1017/9781009325844.

Intergovernmental Panel On Climate Change. *Climate Change 2021 – The Physical Science Basis: Working Group I Contribution to the Sixth Assessment Report of the Intergovernmental Panel on Climate Change.* 1st ed. Cambridge University Press, 2023. https://doi.org/10.1017/9781009157896.

Iyengar, Shanto, Gaurav Sood, and Yphtach Lelkes. "Affect, Not Ideology." *Public Opinion Quarterly* 76, no. 3 (2012): 405–31. https://doi.org/10.1093/poq/nfs038.

Jamieson, Lynn. "Sociologies of Personal Relationships and the Challenge of Climate Change." *Sociology* 54, no. 2 (2020): 219–36. https://doi.org/10.1177/0038038519882599.

Janssen, Marco A., Timothy A. Kohler, and Marten Scheffer. "Sunk-Cost Effects and Vulnerability to Collapse in Ancient Societies." *Current Anthropology* 44, no. 5 (2003): 722–28. https://doi.org/10.1086/379261.

Jaro'ah, Siti, Vania Ardelia, Nurchayati, and Miftakhul Jannah. "Climate Is More Than Just Weather: Gap of Knowledge about Climate Change and Its Psychological Impacts among Indonesian Youth." *Indonesian Journal of Social and Environmental Issues (IJSEI)* 4, no. 2 (2023): 160–70. https://doi.org/10.47540/ijsei.v4i2.1001.

Jasanoff, Sheila. "A New Climate for Society." *Theory, Culture & Society* 27, no. 2–3 (2010): 233–53. https://doi.org/10.1177/0263276409361497.

Jasper, James M. "The Emotions of Protest: Affective and Reactive Emotions in and around Social Movements." *Sociological Forum* 13, no. 3 (1998): 397–424.

Jenkins, Laura. "Why Do All Our Feelings about Politics Matter?" *The British Journal of Politics and International Relations* 20, no. 1 (2018): 191–205. https://doi.org/10.1177/1369148117746917.

Jylhä, Kirsti M., and Nazar Akrami. "Social Dominance Orientation and Climate Change Denial: The Role of Dominance and System Justification." *Personality and Individual Differences* 86 (2015): 108–11. https://doi.org/10.1016/j.paid.2015.05.041.

Kahneman, Daniel. *Thinking, Fast and Slow.* New York, NY: Farrar, Straus and Giroux, 2011.

Kapeller, Marie Lisa, and Georg Jäger. "Threat and Anxiety in the Climate Debate—An Agent-Based Model to Investigate Climate Scepticism and Pro-Environmental Behaviour." *Sustainability* 12, no. 5 (2020): 1823. https://doi.org/10.3390/su12051823.

Karsgaard, Carrie, and Debra Davidson. "Must We Wait for Youth to Speak Out before We Listen? International Youth Perspectives and Climate Change Education." *Educational Review* 75, no. 1 (2023): 74–92. https://doi.org/10.1080/00131911.2021.1905611.

Kemeny, Margaret E., Tara L. Gruenewald, and Sally S. Dickerson. "Shame as the Emotional Response to Threat to the Social Self: Implications for Behavior, Physiology, and Health." *Psychological Inquiry* 15, no. 2 (2004): 153–60.

Kemkes, Robin J., and Sean Akerman. "Contending with the Nature of Climate Change: Phenomenological Interpretations from Northern Wisconsin." *Emotion, Space and Society* 33 (2019): 100614. https://doi.org/10.1016/j.emospa.2019.100614.

Kemp, Luke, Chi Xu, Joanna Depledge, Kristie L. Ebi, Goodwin Gibbins, Timothy A. Kohler, Johan Rockström, et al. "Climate Endgame: Exploring Catastrophic Climate Change Scenarios." *Proceedings of the National Academy of Sciences* 119, no. 34 (2022): e2108146119. https://doi.org/10.1073/pnas.2108146119.

Kemper, Theodore D. *A Social Interactional Theory of Emotions.* New York, NY: Wiley, 1978.

Kimmerer, Robin Wall. *Braiding Sweetgrass: Indigenous Wisdom, Scientific Knowledge and the Teachings of Plants.* First paperback edition. Minneapolis, MN: Milkweed Editions, 2013.

Klein, Naomi. *This Changes Everything: Capitalism vs. the Climate*. Toronto: Vintage Canada, 2015.
Klein, Naomi. *The Shock Doctrine: The Rise of Disaster Capitalism*. New York, NY: Metropolitan Books/Henry Holt and Company, 2007.
Kleres, Jochen, and Åsa Wettergren. "Fear, Hope, Anger, and Guilt in Climate Activism." *Social Movement Studies* 16, no. 5 (2017): 507–19. https://doi.org/10.1080/14742837.2017.1344546.
Knez, I., A. Butler, Å. Ode Sang, E. Ångman, I. Sarlöv-Herlin, and A. Åkerskog. "Before and after a Natural Disaster: Disruption in Emotion Component of Place-Identity and Wellbeing." *Journal of Environmental Psychology* 55 (2018): 11–17. https://doi.org/10.1016/j.jenvp.2017.11.002.
Kozhisseri, Deepa. "'Valli' at the Border: Adivasi Women de-Link from Settler Colonialism Paving Reenchantment of the Forest Commons." *Journal of International Women's Studies* 21, no. 7 (2020): 139–52.
Krange, Olve, Bjørn P. Kaltenborn, and Martin Hultman. "Cool Dudes in Norway: Climate Change Denial among Conservative Norwegian Men." *Environmental Sociology* 5, no. 1 (2019): 1–11. https://doi.org/10.1080/23251042.2018.1488516.
Krznaric, Roman. *Empathy*. Random House, 2014.
Langford, Dale J., Sara E. Crager, Zarrar Shehzad, Shad B. Smith, Susana G. Sotocinal, Jeremy S. Levenstadt, Mona Lisa Chanda, Daniel J. Levitin, and Jeffrey S. Mogil. "Social Modulation of Pain as Evidence for Empathy in Mice." *Science* 312, no. 5872 (2006): 1967–70.
LeBrón, Marisol. "Policing Coraje in the Colony: Toward a Decolonial Feminist Politics of Rage in Puerto Rico." *Signs: Journal of Women in Culture and Society* 46, no. 4 (2021): 801–26.
LeDoux, Joseph. *The Emotional Brain: The Mysterious Underpinnings of Emotional Life*. New York, NY: Simon and Schuster, 1996.
Legerstee, Maria. "The Role of Person and Object in Eliciting Early Imitation." *Journal of Experimental Child Psychology* 51 (1991): 423–33.
Lemke, Thomas. "3. The Risks of Security: Liberalism, Biopolitics, and Fear." In *The Government of Life*, edited by Vanessa Lemm and Miguel Vatter, 59–74. Fordham University Press, 2020. https://doi.org/10.1515/9780823256006-006.
Lenton, Timothy M., David I. A. McKay, Sina Loriani, Jesse F. Abrams, Steven J. Lade, Jonathan F. Donges, Joshua E. Buxton, et al. "The Global Tipping Points Report." Exeter: University of Exeter, 2023.
LeQuesne, Theo. "Petro-Hegemony and the Matrix of Resistance: What Can Standing Rock's Water Protectors Teach Us about Organizing for Climate Justice in the United States?" *Environmental Sociology* 5, no. 2 (2019): 188–206. https://doi.org/10.1080/23251042.2018.1541953.
Letourneau, Angeline M., and Debra Davidson. "Farmer Identities: Facilitating Stability and Change in Agricultural System Transitions." *Environmental Sociology* 8, no. 4 (2022): 459–70. https://doi.org/10.1080/23251042.2022.2064207.
Lidskog, Rolf, Monika Berg, Karin M. Gustafsson, and Erik Löfmarck. "Cold Science Meets Hot Weather: Environmental Threats, Emotional Messages and Scientific Storytelling." *Media and Communication* 8, no. 1 (2020): 118–28. https://doi.org/10.17645/mac.v8i1.2432.
Liévanos, Raoul S. "Air-Toxic Clusters Revisited: Intersectional Environmental Inequalities and Indigenous Deprivation in the U.S. Environmental Protection Agency Regions." *Race and Social Problems* 11, no. 2 (2019): 161–84. https://doi.org/10.1007/s12552-019-09260-5.
Lin, Pei-Ying, Naomi S. Grewal, Christophe Morin, Walter D. Johnson, and Paul J. Zak. "Oxytocin Increases the Influence of Public Service Advertisements." *PLOS One* 8, no. 2 (2013): e56934.

Livingstone, Louise. "Taking Sustainability to Heart—Towards Engaging with Sustainability Issues Through Heart-Centred Thinking." In *Sustainability and the Humanities*, edited by Walter Leal Filho and Adriana Consorte McCrea, 455–67. Cham: Springer International Publishing, 2019. https://doi.org/10.1007/978-3-319-95336-6_26.

Lu, Hang, and Jonathon P. Schuldt. "Exploring the Role of Incidental Emotions in Support for Climate Change Policy." *Climatic Change* 131, no. 4 (2015): 719–26. https://doi.org/10.1007/s10584-015-1443-x.

Luo, Jiayi, and Rongjun Yu. "Follow the Heart or the Head? The Interactive Influence Model of Emotion and Cognition." *Frontiers in Psychology* 6 (2015). https://doi.org/10.3389/fpsyg.2015.00573.

Lutz, Catherine. "Emotions and Feminist Theories." In *Querelles: Jahrbuch für Frauenforschung 2002*, edited by Ingrid Kasten, Gesa Stedman, and Margarete Zimmermann, 104–21. Stuttgart: J.B. Metzler, 2002. https://doi.org/10.1007/978-3-476-02869-3_6.

Lynch, Michael. "We Have Never Been Anti-Science: Reflections on Science Wars and Post-Truth." *Engaging Science, Technology, and Society* 6 (2020): 49–57. https://doi.org/10.17351/ests2020.309.

Markowitz, Ezra M., and Azim F. Shariff. "Climate Change and Moral Judgement." *Nature Climate Change* 2, no. 4 (2012): 243–47. https://doi.org/10.1038/nclimate1378.

Markowitz, Ezra M., Paul Slovic, Daniel Västfjäll, and Sara D. Hodges. "Compassion Fade and the Challenge of Environmental Conservation." *Judgment and Decision Making* 8, no. 4 (2013): 397–406. https://doi.org/10.1017/S193029750000526X.

Martin, Susanne. "On the Persistence of Fear in Late Capitalism: Insights from Modernisation Theories and Affect Theories." *Emotions and Society* 4, no. 3 (2022): 307–22. https://doi.org/10.1332/263169021X16623713200649.

Mayer, Adam P., and E. Keith Smith. "Multidimensional Partisanship Shapes Climate Policy Support and Behaviours." *Nature Climate Change* 13, no. 1 (2023): 32–39. https://doi.org/10.1038/s41558-022-01548-6.

McCaffree, Kevin. "Towards an Integrative Sociological Theory of Empathy." *European Journal of Social Theory* 23, no. 4 (November 2020): 550–70. https://doi.org/10.1177/1368431019890494.

McCoy, Shannon K., and Brenda Major. "Group Identification Moderates Emotional Responses to Perceived Prejudice." *Personality and Social Psychology Bulletin* 29, no. 8 (2003): 1005–17. https://doi.org/10.1177/0146167203253466.

McCright, Aaron M., and Riley E. Dunlap. "Bringing Ideology in: The Conservative White Male Effect on Worry about Environmental Problems in the USA." *Journal of Risk Research* 16, no. 2 (2013): 211–26. https://doi.org/10.1080/13669877.2012.726242.

McGregor, Heather. "Conceptualising Male Violence against Female Partners: Political Implications of Therapeutic Responses." *Australian and New Zealand Journal of Family Therapy* 11, no. 2 (1990): 65–70. https://doi.org/10.1002/j.1467-8438.1990.tb00793.x.

McIntosh, Roderick J., Joseph A. Tainter, and Susan Keech McIntosh, eds. *The Way the Wind Blows: Climate, History, and Human Action*. The Historical Ecology Series. New York: Columbia University Press, 2000.

Mellor, Mary. *Breaking the Boundaries: Towards a Feminist Green Socialism*. Politics/Social Issues. London: Virago Press, 1992.

Merkley, Eric, and Dominik A. Stecula. "Party Cues in the News: Democratic Elites, Republican Backlash, and the Dynamics of Climate Skepticism." *British Journal of Political Science* 51, no. 4 (2021): 1439–56. https://doi.org/10.1017/S0007123420000113.

Meyer, John W., and Brian Rowan. "Institutionalized Organizations: Formal Structure as Myth and Ceremony." *American Journal of Sociology* 83, no. 2 (1977): 340–63.

Miceli, Maria, and Cristiano Castelfranchi. "Hope: The Power of Wish and Possibility." *Theory & Psychology* 20, no. 2 (2010): 251–76. https://doi.org/10.1177/0959354309354393.

Milfont, Taciano L., Isabel Richter, Chris G. Sibley, Marc S. Wilson, and Ronald Fischer. "Environmental Consequences of the Desire to Dominate and Be Superior." *Personality and Social Psychology Bulletin* 39, no. 9 (2013): 1127–38. https://doi.org/10.1177/0146167213490805.

Milman, Oliver. "'Silent Killer': Experts Warn of Record US Deaths from Extreme Heat." *The Guardian*, August 1, 2023, sec. US News.

Milton, Kay. *Loving Nature: Towards an Ecology of Emotion*. London: Routledge, 2002.

Moon, Dawne. "Powerful Emotions: Symbolic Power and the (Productive and Punitive) Force of Collective Feeling." *Theory and Society* 42, no. 3 (2013): 261–94. https://doi.org/10.1007/s11186-013-9190-3.

Moore, Jason W. *Capitalism in the Web of Life: Ecology and the Accumulation of Capital*. 1st ed. New York: Verso, 2015.

Mora, Camilo, Tristan McKenzie, Isabella M. Gaw, Jacqueline M. Dean, Hannah Von Hammerstein, Tabatha A. Knudson, Renee O. Setter, et al. "Over Half of Known Human Pathogenic Diseases Can Be Aggravated by Climate Change." *Nature Climate Change* 12, no. 9 (2022): 869–75. https://doi.org/10.1038/s41558-022-01426-1.

Morris, Brandi S., Polymeros Chrysochou, Jacob Dalgaard Christensen, Jacob L. Orquin, Jorge Barraza, Paul J. Zak, and Panagiotis Mitkidis. "Stories vs. Facts: Triggering Emotion and Action-Taking on Climate Change." *Climatic Change* 154, no. 1–2 (2019): 19–36. https://doi.org/10.1007/s10584-019-02425-6.

Morris, Stephen G. "Empathy and the Liberal-Conservative Political Divide in the U.S." *Journal of Social and Political Psychology* 8, no. 1 (2020): 8–24. https://doi.org/10.5964/jspp.v8i1.1102.

Mortreux, Colette, Jon Barnett, Sergio Jarillo, and Katharine H. Greenaway. "Reducing Personal Climate Anxiety Is Key to Adaptation." *Nature Climate Change* 13, no. 7 (2023): 590–590. https://doi.org/10.1038/s41558-023-01716-2.

Moser, Susanne C. "Communicating Climate Change: History, Challenges, Process and Future Directions." *Wiley Interdisciplinary Reviews Climate Change* 1, no. 1 (2010): 32–53.

Moser, Susanne C., and Lisa Dilling. "Communicating Climate Change: Closing the Science-Action Gap." In *The Oxford Handbook of Climate Change and Society*, edited by John S. Dryzek, Richard B. Norgaard, and David Schlosberg, 161–74. London: Oxford University Press, 2012.

Müller, Birgit. "'To Act upon One's Time ...' From the Impulse to Resist to Global Political Strategy." *Anthropological Theory* 19, no. 1 (2019): 54–73. https://doi.org/10.1177/1463499618792168.

Muscatell, Keely A. "Brains, Bodies, and Social Hierarchies." *Cerebrum*, no. Jan–Feb (2020).

Nakahashi, Wataru, and Hisashi Ohtsuki. "When Is Emotional Contagion Adaptive?" *Journal of Theoretical Biology* 380 (2015): 480–88. https://doi.org/10.1016/j.jtbi.2015.06.014.

Navarro, Oscar, Nathalie Krien, Delphine Rommel, Aurore Deledalle, Colin Lemée, Marie Coquet, Denis Mercier, and Ghozlane Fleury-Bahi. "Coping Strategies Regarding Coastal Flooding Risk in a Context of Climate Change in a French

Caribbean Island." *Environment and Behavior* 53, no. 6 (2021): 636–60. https://doi.org/10.1177/0013916520916253.

Norgaard, Kari Marie. *Salmon and Acorns Feed Our People: Colonialism, Nature, and Social Action.* Nature, Society, and Culture. New Brunswick, NJ: Rutgers University Press, 2019.

Norgaard, Kari Marie. *Living in Denial: Climate Change, Emotions, and Everyday Life.* Cambridge, MA: MIT Press, 2011.

Norgaard, Kari Marie, and Ron Reed. "Emotional Impacts of Environmental Decline: What Can Native Cosmologies Teach Sociology about Emotions and Environmental Justice?" *Theory and Society* 46, no. 6 (2017): 463–95. https://doi.org/10.1007/s11186-017-9302-6.

Nussbaum, Martha. *For Love of Country: Debating the Limits of Patriotism.* Boston: Beacon Press, 1996.

Ogunbode, Charles A., Gisela Böhm, Stuart B. Capstick, Christina Demski, Alexa Spence, and Nicole Tausch. "The Resilience Paradox: Flooding Experience, Coping and Climate Change Mitigation Intentions." *Climate Policy* 19, no. 6 (2019): 703–15. https://doi.org/10.1080/14693062.2018.1560242.

Ogunbode, Charles A., Rouven Doran, and Gisela Böhm. "Individual and Local Flooding Experiences Are Differentially Associated with Subjective Attribution and Climate Change Concern." *Climatic Change* 162, no. 4 (2020): 2243–55. https://doi.org/10.1007/s10584-020-02793-4.

Ogunbode, Charles A., Rouven Doran, Daniel Hanss, Maria Ojala, Katariina Salmela-Aro, Karlijn L. Van Den Broek, Navjot Bhullar, et al. "Climate Anxiety, Wellbeing and pro-Environmental Action: Correlates of Negative Emotional Responses to Climate Change in 32 Countries." *Journal of Environmental Psychology* 84 (2022): 101887. https://doi.org/10.1016/j.jenvp.2022.101887.

Oil Change International. "Planet Wreckers: How 20 Ountires' Oil and Gas Extraction Plans Risk Locking in Climate Chaos." Oil Change International, 2023.

Oksala, Johanna. "Feminism, Capitalism, and Ecology." *Hypatia* 33, no. 2 (2018): 216–34. https://doi.org/10.1111/hypa.12395.

Oladejo, Toheeb Olalekan, Fatai Omeiza Balogun, Usman Abubakar Haruna, Hassan Olayemi Alaka, Joseph Almazan, Musa Saidu Shuaibu, Ibrahim Sheu Adedayo, Zhanerke Ermakhan, Antonio Sarria-Santamerra, and Don Lucero-Prisno Eliseo. "Climate Change in Kazakhstan: Implications to Population Health." *Bulletin of the National Research Centre* 47, no. 1 (2023): 144. https://doi.org/10.1186/s42269-023-01122-w.

Olson, Mancur. *The Logic of Collective Action: Public Goods and the Theory of Groups.* 21. printing. Harvard Economic Studies 124. Cambridge, MA: Harvard University Press, 2003.

Osberghaus, Daniel, and Carina Fugger. "Natural Disasters and Climate Change Beliefs: The Role of Distance and Prior Beliefs." *Global Environmental Change* 74 (2022): 102515. https://doi.org/10.1016/j.gloenvcha.2022.102515.

Pagel, Mark. *Wired for Culture: Origins of the Human Social Mind.* New York, NY: W. W. Norton, 2012.

Pain, Rachel. "Globalized Fear? Towards an Emotional Geopolitics." *Progress in Human Geography* 33, no. 4 (2009): 466–86. https://doi.org/10.1177/0309132508104994.

Panksepp, Jaak. *Affective Neuroscience: The Foundations of Human and Animal Emotions.* Oxford: Oxford University Press, 1998.

Pasca, Laura. "Pride and Guilt as Mediators in the Relationship between Connection to Nature and Pro-Environmental Intention." *Climatic Change* 175, no. 1–2 (2022): 5. https://doi.org/10.1007/s10584-022-03458-0.

Passyn, Kirsten, and Mita Sujan. "Self-Accountability Emotions and Fear Appeals: Motivating Behavior." *Journal of Consumer Research* 32, March (2006): 583–89.

Pavletich, JoAnn. "Emotions, Experience, and Social Control in the Twentieth Century." *Rethinking Marxism* 10, no. 2 (1998): 51–64. https://doi.org/10.1080/08935699808685526.

Pearce, Joshua M., and Richard Parncutt. "Quantifying Global Greenhouse Gas Emissions in Human Deaths to Guide Energy Policy." *Energies* 16, no. 16 (2023): 6074. https://doi.org/10.3390/en16166074.

Pease, Bob. "The Politics of Gendered Emotions: Disrupting Men's Emotional Investment in Privilege." *Australian Journal of Social Issues* 47, no. 1 (2012): 125–42. https://doi.org/10.1002/j.1839-4655.2012.tb00238.x.

Perkins, Helen E. "Measuring Love and Care for Nature." *Journal of Environmental Psychology* 30, no. 4 (2010): 455–63. https://doi.org/10.1016/j.jenvp.2010.05.004.

Perrow, Charles. *Normal Accidents: Living with High-Risk Technologies ; with a New Afterword and a Postscript on the Y2K Problem.* Repr. Princeton, NJ: Princeton University Press, 1999.

Peters, Ellen, and Paul Slovic. "The Springs of Action: Affective and Analytical Information Processing in Choice." *Personality and Social Psychology Bulletin* 26, no. 12 (2000): 1465–75. https://doi.org/10.1177/01461672002612002.

Pfattheicher, Stefan, Claudia Sassenrath, and Simon Schindler. "Feelings for the Suffering of Others and the Environment: Compassion Fosters Proenvironmental Tendencies." *Environment and Behavior* 48, no. 7 (2016): 929–45.

Pfister, Hans-Rüdiger, and Gisela Böhm. "The Multiplicity of Emotions: A Framework of Emotional Functions in Decision Making." *Judgment and Decision Making* 3, no. 1 (2008): 5–17.

Phoenix, Davin L. *The Anger Gap: How Race Shapes Emotion in Politics.* Cambridge: Cambridge University Press, 2019.

Piff, Paul K., Daniel M. Stancato, Stéphane Côté, Rodolfo Mendoza-Denton, and Dacher Keltner. "Higher Social Class Predicts Increased Unethical Behavior." *Proceedings of the National Academy of Sciences* 109, no. 11 (2012): 4086–91.

Piketty, Thomas. *Capital in the Twenty-First Century.* Translated by Arthur Goldhammer. Cambridge, MA, London: The Belknap Press of Harvard University Press, 2017.

Pinchoff, Jessie, Ricardo Regules, Ana C. Gomez-Ugarte, Tara F. Abularrage, and Ietza Bojorquez-Chapela. "Coping with Climate Change: The Role of Climate Related Stressors in Affecting the Mental Health of Young People in Mexico." Edited by Kévin Jean. *PLOS Global Public Health* 3, no. 9 (2023): e0002219. https://doi.org/10.1371/journal.pgph.0002219.

Pratto, Felicia, Jim Sidanius, Lisa M. Stallworth, and Bertram F. Malle. "Social Dominance Orientation: A Personality Variable Predicting Social and Political Attitudes." *Journal of Personality and Social Psychology* 67, no. 4 (1994): 741–63. https://doi.org/10.1037/0022-3514.67.4.741.

Prinz, Wolfgang. "Perception and Action Planning." *European Journal of Cognitive Psychology* 9 (1997): 129–54.

Pulé, Paul M., and Martin Hultman, eds. *Men, Masculinities, and Earth: Contending with the (m)Anthropocene.* Cham: Springer International Publishing, 2021. https://doi.org/10.1007/978-3-030-54486-7.

Pulé, Paul M., Martin Hultman, and Angelica Wägstrom. "Discussions at the Table." In *Men, Masculinities, and Earth: Contending with the (m)Anthropocene*, edited by Paul M. Pulé and Martin Hultman, 17–101. Cham, Switzerland: Palgrave Macmillan, 2021.

Rached, Melissa A., Ahmed Hankir, and Rashid Zaman. "Emotional Abuse in Women and Girls Mediated by Patriarchal Upbringing and Its Impact on Sexism

and Mental Health: A Narrative Review." *Psychiatria Danubina* 33, no. Suppl 11 (2021): 137–44.
Radel, Claudia, Lindsey Carte, Richard L. Johnson, and Birgit Schmook. "Emotions and Gendered Experiences of Livelihood Migration: Memos from Nicaragua and Guatemala: GPC Special Issue: 'Towards Feminist Geographies of Livelihoods.'" *Gender, Place & Culture* (2023): 1–10. https://doi.org/10.1080/0966369X.2023.2249251.
Reser, Joseph P., and Janet K. Swim. "Adapting to and Coping with the Threat and Impacts of Climate Change." *American Psychologist* 66, no. 4 (2011): 277–89. https://doi.org/10.1037/a0023412.
Richerson, Peter J., and Robert Boyd. *Not by Genes Alone: How Culture Transformed Human Evolution.* Chicago, IL: University of Chicago Press, 2005.
Richez, Emmanuelle, Vincent Raynauld, Abunya Agi, and Arief B. Kartolo. "Unpacking the Political Effects of Social Movements With a Strong Digital Component: The Case of #IdleNoMore in Canada." *Social Media + Society* 6, no. 2 (2020): 205630512091558. https://doi.org/10.1177/2056305120915588.
Rico, Guillem, Marc Guinjoan, and Eva Anduiza. "The Emotional Underpinnings of Populism: How Anger and Fear Affect Populist Attitudes." *Swiss Political Science Review* 23, no. 4 (2017): 444–61. https://doi.org/10.1111/spsr.12261.
Rikani, Albano, Christian Otto, Anders Levermann, and Jacob Schewe. "More People Too Poor to Move: Divergent Effects of Climate Change on Global Migration Patterns." *Environmental Research Letters* 18, no. 2 (2023): 024006. https://doi.org/10.1088/1748-9326/aca6fe.
Rimé, Bernard. "Emotions at the Service of Cultural Construction." *Emotion Review* 12, no. 2 (2020): 65–78. https://doi.org/10.1177/1754073919876036.
Rinscheid, Adrian, Silvia Pianta, and Elke U. Weber. "What Shapes Public Support for Climate Change Mitigation Policies? The Role of Descriptive Social Norms and Elite Cues." *Behavioural Public Policy* 5, no. 4 (2021): 503–27. https://doi.org/10.1017/bpp.2020.43.
Rising, James, Marco Tedesco, Franziska Piontek, and David A. Stainforth. "The Missing Risks of Climate Change." *Nature* 610, no. 7933 (2022): 643–51. https://doi.org/10.1038/s41586-022-05243-6.
Rizzolatti, Giacomo, and Corrado Sinigaglia. "Mirror Neurons and Motor Intentionality." *Functional Neurology* 22, no. 4 (2007): 205–10.
Robinson, Cedric J. *On Racial Capitalism, Black Internationalism, and Cultures of Resistance.* London: Pluto Press, 2019.
Rochat, Philippe, and Tricia Striano. "Social Cognitive Development in the First Year." In *Early Social Cognition: Understanding Others in the First Months of Life*, edited by Philippe Rochat, 3–34. Mahwah, NJ: Erlbaum, 1999.
Rollert, John Paul. "The Wages of Intimate and Anonymous Capitalism." *Society* 55, no. 3 (2018): 271–79. https://doi.org/10.1007/s12115-018-0250-1.
Rolls, Edmund. *Emotion and Decision Making Explained.* London: Oxford University Press, 2014.
Romanello, Marina, Claudia Di Napoli, Paul Drummond, Carole Green, Harry Kennard, Pete Lampard, Daniel Scamman, et al. "The 2022 Report of the Lancet Countdown on Health and Climate Change: Health at the Mercy of Fossil Fuels." *The Lancet* 400, no. 10363 (2022): 1619–54. https://doi.org/10.1016/S0140-6736(22)01540-9.
Ruiz-Junco, Natalia. "Advancing the Sociology of Empathy: A Proposal." *Symbolic Interaction* 40, no. 3 (2017): 414–35. https://doi.org/10.1002/symb.306.
Russell, James. *Agency and Its Role in Mental Development.* London: Psychology Press, 1996.

186 Works Cited

Rutherford, Jonathan. *MEN'S SILENCES Predicaments in Masculinity*. New York, NY: Routledge, 2023.
Sabherwal, Anandita, Adam R. Pearson, and Gregg Sparkman. "Anger Consensus Messaging Can Enhance Expectations for Collective Action and Support for Climate Mitigation." *Journal of Environmental Psychology* 76 (2021): 101640. https://doi.org/10.1016/j.jenvp.2021.101640.
Salmela, Mikko, and Christian Von Scheve. "Emotional Roots of Right-Wing Political Populism." *Social Science Information* 56, no. 4 (2017): 567–95. https://doi.org/10.1177/0539018417734419.
Sangervo, Julia, Kirsti M. Jylhä, and Panu Pihkala. "Climate Anxiety: Conceptual Considerations, and Connections with Climate Hope and Action." *Global Environmental Change* 76 (2022): 102569. https://doi.org/10.1016/j.gloenvcha.2022.102569.
Sapolsky, Robert M. *Behave: The Biology of Humans at Our Best and Worst*. New York, NY: Penguin Books, 2017.
Schild, Verónica. "Feminisms, the Environment and Capitalism: On the Necessary Ecological Dimension of a Critical Latin American Feminism." *Journal of International Women's Studies* 20, no. 6 (2019): 23–43.
Schneider, Claudia R., and Sander Van Der Linden. "An Emotional Road to Sustainability: How Affective Science Can Support pro-Climate Action." *Emotion Review* 15, no. 4 (2023): 284–88. https://doi.org/10.1177/17540739231193742.
Schneider, Claudia R., Lisa Zaval, Elke U. Weber, and Ezra M. Markowitz. "The Influence of Anticipated Pride and Guilt on Pro-Environmental Decision Making." Edited by Brock Bastian. *PLOS ONE* 12, no. 11 (2017): e0188781. https://doi.org/10.1371/journal.pone.0188781.
Schwartz, Barry. "How Is History Possible? Georg Simmel on Empathy and Realism." *Journal of Classical Sociology* 17, no. 3 (2017): 213–37. https://doi.org/10.1177/1468795X17717877.
Segal, Elizabeth. *Social Empathy: The Art of Understanding Others*. New York, NY: Columbia University Press, 2018.
Senande-Rivera, Martín, Damián Insua-Costa, and Gonzalo Miguez-Macho. "Spatial and Temporal Expansion of Global Wildland Fire Activity in Response to Climate Change." *Nature Communications* 13, no. 1 (2022): 1208. https://doi.org/10.1038/s41467-022-28835-2.
Seo, Minjae, Shiyu Yang, and Sean M. Laurent. "No One Is an Island: Awe Encourages Global Citizenship Identification." *Emotion* 23, no. 3 (2023): 601–12. https://doi.org/10.1037/emo0001160.
Seremetakis, Constantina Nadia. *The Last Word: Women, Death, and Divination in Inner Mani*. Nachdr. Chicago: University of Chicago Press, 1997.
Shao, Lei, and Guoliang Yu. "Media Coverage of Climate Change, Eco-Anxiety and pro-Environmental Behavior: Experimental Evidence and the Resilience Paradox." *Journal of Environmental Psychology* 91 (2023): 102130. https://doi.org/10.1016/j.jenvp.2023.102130.
Shaw, Duncan, Judy Scully, and Tom Hart. "The Paradox of Social Resilience: How Cognitive Strategies and Coping Mechanisms Attenuate and Accentuate Resilience." *Global Environmental Change* 25 (2014): 194–203. https://doi.org/10.1016/j.gloenvcha.2014.01.006.
Shepard, Roger N. "Ecological Constraints on Internal Representation: Resonant Kinematics of Perceiving, Imagining, Thinking, and Dreaming." *Psychological Review* 91 (1984): 417–77.
Shepherd, Steven, and Aaron C. Kay. "On the Perpetuation of Ignorance: System Dependence, System Justification, and the Motivated Avoidance of Sociopolitical

Information." *Journal of Personality and Social Psychology* 102, no. 2 (2012): 264–80. https://doi.org/10.1037/a0026272.

Shipley, Nathan J., and Carena J. Van Riper. "Pride and Guilt Predict Pro-Environmental Behavior: A Meta-Analysis of Correlational and Experimental Evidence." *Journal of Environmental Psychology* 79 (2022): 101753. https://doi.org/10.1016/j.jenvp.2021.101753.

Shoemaker, Pamela J. "Hardwired for News: Using Biological and Cultural Evolution to Explain the Surveillance Function." *Journal of Communication* 46, no. 3 (1996): 32–47. https://doi.org/10.1111/j.1460-2466.1996.tb01487.x.

Simas, Elizabeth N., Scott Clifford, and Justin H. Kirkland. "How Empathic Concern Fuels Political Polarization." *American Political Science Review* 114, no. 1 (2020): 258–69. https://doi.org/10.1017/S0003055419000534.

Skurka, Chris, Jessica Gall Myrick, and Yin Yang. "Fanning the Flames or Burning Out? Testing Competing Hypotheses about Repeated Exposure to Threatening Climate Change Messages." *Climatic Change* 176, no. 5 (2023): 52. https://doi.org/10.1007/s10584-023-03539-8.

Skurka, Christofer, Jeff Niederdeppe, Rainer Romero-Canyas, and David Acup. "Pathways of Influence in Emotional Appeals: Benefits and Tradeoffs of Using Fear or Humor to Promote Climate Change-Related Intentions and Risk Perceptions." *Journal of Communication* 68, no. 1 (2018): 169–93. https://doi.org/10.1093/joc/jqx008.

Solnit, Rebecca. *A Paradise Built in Hell: The Extraordinary Communities That Arise in Disasters.* New York: Viking, 2009.

Sparkman, Gregg, Nathan Geiger, and Elke U. Weber. "Americans Experience a False Social Reality by Underestimating Popular Climate Policy Support by Nearly Half." *Nature Communications* 13, no. 1 (2022): 4779. https://doi.org/10.1038/s41467-022-32412-y.

Spence, Alexa, and Charles A. Ogunbode. "Angry Politics Fails the Climate." *Nature Climate Change* 13, no. 1 (2023): 13–14. https://doi.org/10.1038/s41558-022-01567-3.

Spence, Alexa, Wouter Poortinga, and Nick Pidgeon. "The Psychological Distance of Climate Change." *Risk Analysis* 32, no. 6 (2012): 957–72. https://doi.org/10.1111/j.1539-6924.2011.01695.x.

Stanley, Samantha K., and Marc S. Wilson. "Meta-Analysing the Association between Social Dominance Orientation, Authoritarianism, and Attitudes on the Environment and Climate Change." *Journal of Environmental Psychology* 61 (2019): 46–56. https://doi.org/10.1016/j.jenvp.2018.12.002.

Stefano, George B. "Cognition Regulated by Emotional Decision Making." *Medical Science Monitor Basic Research* 22 (2016): 1–5.

Stein, Edith. *On the Problem of Empathy: The Collected Works of Edith Stein.* Washington: ICS Publications, 1989.

Stengers, Isabelle. *In Catastrophic Times: Resisting the Coming Barbarism.* Translated by Andrew Goffey. GB: Open Humanities Press, Licensed under Creative Commons, 2015. https://doi.org/10.14619/016.

Stephenson, Wen. "Against Climate Barbarism: A Conversation with Naomi Klein." *Los Angeles Review of Books (blog),* September 30, 2019. https://lareviewofbooks.org/article/against-climate-barbarism-a-conversation-with-naomi-klein/.

Stern, Paul C. "Toward a Coherent Theory of Environmentally Significant Behavior." *Journal of Social Issues* 56, no. 3 (2000): 407–24.

Stets, Jan E., and Michael J. Carter. "A Theory of the Self for the Sociology of Morality." *American Sociological Review* 77, no. 1 (2012): 120–40. https://doi.org/10.1177/0003122411433762.

Stewart, Alexander J., Joshua B. Plotkin, and Nolan McCarty. "Inequality, Identity, and Partisanship: How Redistribution Can Stem the Tide of Mass Polarization." *Proceedings of the National Academy of Sciences* 118, no. 50 (2021): e2102140118.

Stollberg, Janine, and Eva Jonas. "Existential Threat as a Challenge for Individual and Collective Engagement: Climate Change and the Motivation to Act." *Current Opinion in Psychology* 42 (2021): 145–50. https://doi.org/10.1016/j.copsyc.2021.10.004.

Sugerman, Eli R., Ye Li, and Eric J. Johnson. "Local Warming Is Real: A Meta-Analysis of the Effect of Recent Temperature on Climate Change Beliefs." *Current Opinion in Behavioral Sciences* 42 (2021): 121–26. https://doi.org/10.1016/j.cobeha.2021.04.015.

Summers-Effler, Erika, Justin Van Ness, and Christopher Hausmann. "Peeking in the Black Box: Studying, Theorizing, and Representing the Micro-Foundations of Day-to-Day Interactions." *Journal of Contemporary Ethnography* 44, no. 4 (2015): 450–79. https://doi.org/10.1177/0891241614545880.

Swim, Janet K., Rosemary Aviste, Michael L. Lengieza, and Carlie J. Fasano. "OK Boomer: A Decade of Generational Differences in Feelings about Climate Change." *Global Environmental Change* 73 (2022): 102479. https://doi.org/10.1016/j.gloenvcha.2022.102479.

Szanto, Thomas. "Collective Emotions, Normativity, and Empathy: A Steinian Account." *Human Studies* 38, no. 4 (2015): 503–27. https://doi.org/10.1007/s10746-015-9350-8.

Szanto, Thomas, and Dermot Moran. "Introduction: Empathy and Collective Intentionality—The Social Philosophy of Edith Stein." *Human Studies* 38, no. 4 (2015): 445–61. https://doi.org/10.1007/s10746-015-9363-3.

Sznycer, Daniel, John Tooby, Leda Cosmides, Roni Porat, Shaul Shalvi, and Eran Halperin. "Shame Closely Tracks the Threat of Devaluation by Others, Even across Cultures." *Proceedings of the National Academy of Sciences* 113, no. 10 (2016): 2625–30. https://doi.org/10.1073/pnas.1514699113.

Taleb, Nassim Nicholas. *The Black Swan: The Impact of the Highly Improbable*. 2nd ed. New York: Random House Trade Paperbacks, 2010.

Tam, Kim-Pong. "Dispositional Empathy with Nature." *Journal of Environmental Psychology* 35 (2013): 92–104. https://doi.org/10.1016/j.jenvp.2013.05.004.

Tam, Kim-Pong, Hoi-Wing Chan, and Susan Clayton. "Climate Change Anxiety in China, India, Japan, and the United States." *Journal of Environmental Psychology* 87 (2023): 101991. https://doi.org/10.1016/j.jenvp.2023.101991.

Tang, Yi-Ting, and Weng-Tink Chooi. "From Concern to Action: The Role of Psychological Distance in Attitude towards Environmental Issues." *Current Psychology* 42, no. 30 (2023): 26570–86. https://doi.org/10.1007/s12144-022-03774-9.

Thiermann, Ute B., and William R. Sheate. "Motivating Individuals for Social Transition: The 2-Pathway Model and Experiential Strategies for pro-Environmental Behaviour." *Ecological Economics* 174 (2020): 106668. https://doi.org/10.1016/j.ecolecon.2020.106668.

Toivonen, Heidi. "Themes of Climate Change Agency: A Qualitative Study on How People Construct Agency in Relation to Climate Change." *Humanities and Social Sciences Communications* 9, no. 1 (2022): 102. https://doi.org/10.1057/s41599-022-01111-w.

Tokita, Christopher K., Andrew M. Guess, and Corina E. Tarnita. "Polarized Information Ecosystems Can Reorganize Social Networks via Information Cascades." *Proceedings of the National Academy of Sciences* 118, no. 50 (2021): e2102147118.

Törnberg, Petter. "How Digital Media Drive Affective Polarization through Partisan Sorting." *Proceedings of the National Academy of Sciences* 119, no. 42 (2022): e2207159119. https://doi.org/10.1073/pnas.2207159119.

Traister, Rebecca. *Good and Mad: The Revolutionary Power of Women's Anger.* New York, NY: S&S/ Marysue Rucci Books, 2023.

Trevarthen, Colwyn, and Kenneth J. Aitken. "Infant Intersubjectivity: Research, Theory, and Clinical Applications." *Journal of Child Psychology and Psychiatry* 42, no. 1 (2001): 3–48. https://doi.org/10.1111/1469-7610.00701.

Turner, Jonathan H. *On Human Nature: The Biology and Sociology of What Made Us Human.* London: Routledge, 2021.

Turner, Jonathan H. *Human Emotions: A Sociological Theory.* 1. publ. London: Routledge, 2007.

Turner, Jonathan H. "Self, Emotions, and Extreme Violence: Extending Symbolic Interactionist Theorizing." *Symbolic Interaction* 30, no. 4 (2007): 501–30. https://doi.org/10.1525/si.2007.30.4.501.

Turner, Jonathan H. *Human Institutions: A Theory of Societal Evolution.* Oxford: Rowman and Littlefield, 2003.

Turner, Jonathan H. *On the Origins of Human Emotions.* Redwood City, CA: Stanford University Press, 2000.

Turner, Jonathan H., and Jan E. Stets. "Sociological Theories of Human Emotions." *Annual Review of Sociology* 32, no. 1 (2006): 25–52. https://doi.org/10.1146/annurev.soc.32.061604.123130.

Tversky, Amos, and Daniel Kahneman. "Judgment under Uncertainty: Heuristics and Biases." *Science* 185, no. 4157 (1974): 1124–31.

Uenal, Fatih, Jim Sidanius, Jon Roozenbeek, and Sander Van Der Linden. "Climate Change Threats Increase Modern Racism as a Function of Social Dominance Orientation and Ingroup Identification." *Journal of Experimental Social Psychology* 97 (2021): 104228. https://doi.org/10.1016/j.jesp.2021.104228.

Van Boven, Leaf, Joanne Kane, A. Peter McGraw, and Jeannette Dale. "Feeling Close: Emotional Intensity Reduces Perceived Psychological Distance." *Journal of Personality and Social Psychology* 98, no. 6 (2010): 872–85. https://doi.org/10.1037/a0019262.

Van Valkengoed, Anne M., Linda Steg, and Goda Perlaviciute. "The Psychological Distance of Climate Change Is Overestimated." *One Earth* 6, no. 4 (2023): 362–91. https://doi.org/10.1016/j.oneear.2023.03.006.

Van Wyk, Hannah, Joseph N. S. Eisenberg, and Andrew F. Brouwer. "Long-Term Projections of the Impacts of Warming Temperatures on Zika and Dengue Risk in Four Brazilian Cities Using a Temperature-Dependent Basic Reproduction Number." Edited by Tomas Leon. *PLOS Neglected Tropical Diseases* 17, no. 4 (2023): e0010839. https://doi.org/10.1371/journal.pntd.0010839.

Veldman, Robin G., Dara M. Wald, Sarah B. Mills, and David A. M. Peterson. "Who Are American Evangelical Protestants and Why Do They Matter for US Climate Policy?" *WIREs Climate Change* 12, no. 2 (2021): e693. https://doi.org/10.1002/wcc.693.

Vercammen, Ans, Britt Wray, Yoshika S. Crider, Gary Belkin, and Emma Lawrance. "Eco-Anxiety and the Influence of Climate Change on Future Planning Is Greater for Young US Residents with Direct Exposure to Climate Impacts." Preprint. In Review, May 24, 2023. https://doi.org/10.21203/rs.3.rs-2698675/v1.

Verlie, Blanche, Emily Clark, Tamara Jarrett, and Emma Supriyono. "Educators' Experiences and Strategies for Responding to Ecological Distress." *Australian Journal of Environmental Education* 37, no. 2 (2021): 132–46. https://doi.org/10.1017/aee.2020.34.

Verweij, Marco, Timothy J. Senior, Juan F. Domínguez D., and Robert Turner. "Emotion, Rationality, and Decision-Making: How to Link Affective and Social Neuroscience with Social Theory." *Frontiers in Neuroscience* 9 (2015): 332. https://doi.org/10.3389/fnins.2015.00332.

Vezzali, Loris, Alessia Cadamuro, Annalisa Versari, Dino Giovannini, and Elena Trifiletti. "Feeling like a Group after a Natural Disaster: Common Ingroup Identity and Relations with Outgroup Victims among Majority and Minority Young Children." *British Journal of Social Psychology* 54, no. 3 (2015): 519–38. https://doi.org/10.1111/bjso.12091.

Vollhardt, Johanna R., and Ervin Staub. "Inclusive Altruism Born of Suffering: The Relationship between Adversity and Prosocial Attitudes and Behavior toward Disadvantaged Outgroups." *American Journal of Orthopsychiatry* 81, no. 3 (2011): 307–15. https://doi.org/10.1111/j.1939-0025.2011.01099.x.

Voronov, Maxim, and Russ Vince. "Integrating Emotions into the Analysis of Institutional Work." *Academy of Management Review* 37, no. 1 (2012): 58–81. https://doi.org/10.5465/amr.2010.0247.

Voronov, Maxim, and Klaus Weber. "The Heart of Institutions: Emotional Competence and Institutional Actorhood." *Academy of Management Review* 41, no. 3 (2016): 456–78. https://doi.org/10.5465/amr.2013.0458.

Walby, Sylvia. *Theorizing Patriarchy*. Reprinted. Oxford: Blackwell, 1997.

Wallace-Wells, David. *The Uninhabitable Earth: Life after Warming*. 1st ed. New York: Tim Duggan Books, 2019.

Walton, Chris, Adrian Coyle, and Evanthia Lyons. "Death and Football: An Analysis of Men's Talk about Emotions." *British Journal of Social Psychology* 43, no. 3 (2004): 401–16. https://doi.org/10.1348/0144666042038024.

Wang, Changcheng, Liuna Geng, and Julián D. Rodríguez-Casallas. "How and When Higher Climate Change Risk Perception Promotes Less Climate Change Inaction." *Journal of Cleaner Production* 321 (2021): 128952. https://doi.org/10.1016/j.jclepro.2021.128952.

Wang, Susie, Zoe Leviston, Mark Hurlstone, Carmen Lawrence, and Iain Walker. "Emotions Predict Policy Support: Why It Matters How People Feel about Climate Change." *Global Environmental Change* 50 (2018): 25–40. https://doi.org/10.1016/j.gloenvcha.2018.03.002.

Watts, Vanessa. "Indigenous Place-Thought & Agency amongst Humans and Non-Humans (First Woman and Sky Woman Go on a European World Tour!)." *Decolonization* 2, no. 1 (2013): 20–34.

Webb, Thomas L., Betty P. I. Chang, and Yael Benn. "'The Ostrich Problem': Motivated Avoidance or Rejection of Information About Goal Progress." *Social and Personality Psychology Compass* 7, no. 11 (2013): 794–807. https://doi.org/10.1111/spc3.12071.

Weber, Elke U. "Seeing Is Believing: Understanding & Aiding Human Responses to Global Climate Change." *Daedalus* 149, no. 4 (2020): 139–50. https://doi.org/10.1162/daed_a_01823.

Webster, Steven W. *American Rage: How Anger Shapes Our Politics*. Cambridge, New York, NY: Cambridge University Press, 2020.

Webster, Steven W., and Bethany Albertson. "Emotion and Politics: Noncognitive Psychological Biases in Public Opinion." *Annual Review of Political Science* 25, no. 1 (2022): 401–18. https://doi.org/10.1146/annurev-polisci-051120-105353.

Weyher, L. Frank. "Re-Reading Sociology via the Emotions: Karl Marx's Theory of Human Nature and Estrangement." *Sociological Perspectives* 55, no. 2 (2012): 341–63. https://doi.org/10.1525/sop.2012.55.2.341.

White, R. K. *Fearful Warriors: A Psychological Profile of U.S.-Soviet Relations*. New York: Free Press, 1984.

Whitmarsh, Lorraine, Lois Player, Angelica Jiongco, Melissa James, Marc Williams, Elizabeth Marks, and Patrick Kennedy-Williams. "Climate Anxiety: What Predicts It and How Is It Related to Climate Action?" *Journal of Environmental Psychology* 83 (2022): 101866. https://doi.org/10.1016/j.jenvp.2022.101866.

Williamson, Clare, Cameron McCordic, and Brent Doberstein. "The Compounding Impacts of Cyclone Idai and Their Implications for Urban Inequality." *International Journal of Disaster Risk Reduction* 86 (2023): 103526. https://doi.org/10.1016/j.ijdrr.2023.103526.

Wilson, Helen F. "Discomfort: Transformative Encounters and Social Change." *Emotion, Space and Society* 37 (2020): 100681. https://doi.org/10.1016/j.emospa.2020.100681.

Witte, Kim, and Mike Allen. "A Meta-Analysis of Fear Appeals: Implications for Effective Public Health Campaigns." *Health Education & Behavior* 27, no. 5 (2000): 591–615.

Wolf, Johanna, and Susanne C. Moser. "Individual Understandings, Perceptions, and Engagement with Climate Change: Insights from In-depth Studies across the World." *WIREs Climate Change* 2, no. 4 (2011): 547–69. https://doi.org/10.1002/wcc.120.

Wolsko, Christopher, Hector Ariceaga, and Jesse Seiden. "Red, White, and Blue Enough to Be Green: Effects of Moral Framing on Climate Change Attitudes and Conservation Behaviors." *Journal of Experimental Social Psychology* 65 (2016): 7–19. https://doi.org/10.1016/j.jesp.2016.02.005.

Wooten, Tom. *We Shall Not Be Moved: Rebuilding Home in the Wake of Katrina.* Boston: Beacon Press, 2012.

Wullenkord, Marlis C., Josephine Tröger, Karen R. S. Hamann, Laura S. Loy, and Gerhard Reese. "Anxiety and Climate Change: A Validation of the Climate Anxiety Scale in a German-Speaking Quota Sample and an Investigation of Psychological Correlates." *Climatic Change* 168, no. 3–4 (2021): 20. https://doi.org/10.1007/s10584-021-03234-6.

Wynne, Brian. "Strange Weather, Again." *Theory, Culture & Society* 27, no. 2–3 (2010): 289–305. https://doi.org/10.1177/0263276410361499.

Yeager, Beth Anne. "Collective Shame in Climate Denial: An Ecopsychological Undertaking." *Ecopsychology* (2023). https://doi.org/10.1089/eco.2023.0002.

Zacher, Hannes, and Cort W. Rudolph. "Environmental Knowledge Is Inversely Associated with Climate Change Anxiety." *Climatic Change* 176, no. 4 (2023): 32. https://doi.org/10.1007/s10584-023-03518-z.

Zanocco, Chad, Hilary Boudet, Roberta Nilson, and June Flora. "Personal Harm and Support for Climate Change Mitigation Policies: Evidence from 10 U.S. Communities Impacted by Extreme Weather." *Global Environmental Change* 59 (2019): 101984. https://doi.org/10.1016/j.gloenvcha.2019.101984.

Zanocco, Chad, June Flora, and Hilary Boudet. "Disparities in Self-Reported Extreme Weather Impacts by Race, Ethnicity, and Income in the United States." Edited by Alessandra Giannini. *PLOS Climate* 1, no. 6 (2022): e0000026. https://doi.org/10.1371/journal.pclm.0000026.

Zaremba, D., M. Kulesza, A. M. Herman, M. Marczak, B. Kossowski, M. Budziszewska, J. M. Michałowski, C. A. Klöckner, A. Marchewka, and M. Wierzba. "A Wise Person Plants a Tree a Day before the End of the World: Coping with the Emotional Experience of Climate Change in Poland." *Current Psychology* 42, no. 31 (2023): 27167–85. https://doi.org/10.1007/s12144-022-03807-3.

Index

accountability 7, 34, 137, 147, 162
activist/ism 120, 136, 139, 140, 147
adaptive capacity 19, 26
affective science/s 8, 59, 62, 82
agency 3–4, 8, 10, 31, 40, 43–4, 52, 55, 58, 68, 74, 76, 85, 94–5, 98, 100, 107, 120, 131–3, 137–9, 144–6, 150–1, 155, 158, 161
anxiety 90, 107, 114, 120–3, 135, 148, 151, 155–6
Atlantic Meridional Overturning Circulation (AMOC) 24

belonging 10–11, 28, 41, 60, 65, 68–9, 75–6, 82, 86–8, 92, 109, 124, 131, 137, 139–40, 156–7, 160, 162

capitalism 4, 9, 35, 40, 45, 58, 83–96, 98, 100, 112, 114–15
civil society 8, 13, 33, 87
CO_2 112, 113
collective action 4, 5, 7–9, 31, 48, 55, 58, 65, 67, 76, 100, 104, 130, 131, 134, 139, 148, 150, 156, 160, 163
colonial/ism 34, 39, 47, 83, 85, 91–100, 114, 118, 139, 157
communication 67, 75, 86, 107
community 25, 27, 45, 69, 86, 87, 99, 100, 110, 118, 124, 134, 136, 138, 140–3, 146, 158–60, 164
complexity 10, 24–5, 49, 51, 58, 62, 137, 144–6
conservative (ideologies) 74, 106, 115–17, 119, 138, 156, 160
cooperation 14, 17, 25, 35, 55, 61, 64–5, 68–9, 73, 91, 154, 157–62

democracy/ies 33, 42, 45, 73, 113, 159–61
denial 35, 72, 105, 107, 111, 113–20, 125, 135, 140, 156, 165

determinism/ist/istic 5, 47, 59, 64, 146, 148
disaster capitalism 26
disaster/s 8, 13, 25–7, 29, 32, 47, 106, 144
disinformation 89, 92, 113

efficacy 66, 72, 73, 85, 91, 100, 107, 123–4, 133, 134, 138–9, 150–1, 156–7, 160, 162, 165
elite/s 2, 3, 19, 49, 52, 116
emissions 6, 14–17, 19, 33, 34, 40, 44, 123–4, 138, 143
empathy/ic 6, 9–11, 62, 65–7, 69–74, 77, 89, 91, 92, 94, 99, 109, 112, 118, 137–9, 142–4, 157
extreme events 13, 20–2, 25–7, 29–30, 47, 106, 122–4, 143, 148

food (security, systems) 22–3, 28, 29, 34, 35, 38, 41, 47, 48, 53, 97, 98, 108, 110, 113, 124, 134, 157, 162
fossil fuels 3, 6, 13, 16–18, 33, 53, 113, 114, 145, 162, 165

gender 7, 59, 68, 73, 83, 92–5, 116, 118, 133, 138
global warming 1–3, 6–7, 9–10, 15–20, 22–5, 27–31, 33, 39–40, 44, 48, 54, 62, 90, 96, 105, 106, 109, 111–16, 119, 121–3, 131, 134–7, 146, 154, 155, 159, 161
greenhouse gas/es 14–15, 18, 40, 124, 138, 162
guilt/y 9, 42, 65–9, 109, 114, 123, 135, 137–8, 140–2, 151

health 23, 25, 26, 29–30, 32, 47, 48, 61, 73, 85, 89, 91, 97, 99, 123, 156
Hochschild, Arlie 89, 142

Index

identity/ies 2, 7, 10, 41, 42, 44, 53, 58, 66, 71, 72, 75–6, 83, 87, 90, 92, 99, 100, 109, 113–19, 139–42, 145, 147, 156, 159
implicatory denial 10
inaction 3, 7, 9–11, 44, 104–5, 107, 111, 124–5, 131, 143, 150–1, 156, 157, 164, 165
indigenous 12, 26, 29, 43, 46, 54, 84–7, 96–100, 119, 139, 143, 148, 157, 164
individualism 4, 83
(in)equity 9, 19, 26, 73, 82, 91, 100, 110, 138, 158, 164
infrastructure 3, 4, 19, 23
in-groups 75, 143
institution/s/al 9, 16, 32–3, 40–5, 47–8, 50–4, 63, 66, 74, 82–6, 99, 100, 107, 110, 124, 131–2, 156, 161
Intergovernmental Panel for Climate Change (IPCC) 15, 16, 18, 19, 22, 28
internet 4, 45, 46, 129
intersectional/ality 59, 84–6, 90, 100, 109, 131, 133, 158

Klein, Naomi 26, 33

labour 29, 40, 54, 89, 92, 93, 95, 96, 118, 155
legitimacy 31, 33, 48, 50, 54, 88, 93, 160
lifestyle 10, 53, 90, 113–15, 121, 130, 133–4, 146

(mal)adaptation/ive 5, 19, 23, 26, 32, 44, 48–51, 54, 64–5, 121–2, 124, 130–3, 154, 162
marginal/ized/ization 26, 73, 85–7, 91, 99, 114, 116, 123, 133
masculinity/ies/ist 35, 94, 118–20
media 3, 9, 12, 28, 73, 75, 88, 92, 105, 107, 110, 141, 147
mental health 25, 26, 29, 30, 61, 73, 85, 89, 91, 123, 156
migration 23, 27–8, 49, 135

neoliberal/ism 6, 8, 13, 86, 93, 114
normalization 35, 119
norms 41–2, 50, 60, 69, 74, 75, 83–5, 93, 94, 110, 119, 137–41, 146, 150

out-groups 33, 72–4, 76, 115, 116, 119, 143

patriarchy 4, 9, 40, 83, 85, 92–7, 100, 118
polycrises 33

population 19, 21–3, 34, 59, 85, 113
pride 9, 10, 42, 43, 62, 65, 67–9, 85, 94, 100, 137, 139–42
pro-climate action 107, 121–2, 130–4, 137–44, 149–51, 156–7, 161–4

reflexive/ivity 3, 4, 6, 11, 49, 52–5, 61, 63, 66, 77, 85, 90, 131–3, 139, 146–8, 155–8, 160, 162, 165
renewable energy 15, 17, 18, 35, 48
Representative Concentration Pathways 14
resilience 26, 27, 49–52, 111
right-wing (extremsism) 2, 28, 33, 91, 113, 118, 159
risk 21, 24, 25, 28, 48, 50, 118; assessment 18; managers/ment 31, 147; perception 9

shame 6, 9, 10, 42, 62, 65–9, 75, 85, 89, 92, 94, 95, 98, 109, 115, 116, 120, 139–42, 151, 165
social capital 40, 48
sociology 7, 8, 40, 47, 63, 64, 115
(social) structure/al 2, 4–11, 19, 32, 40–4, 48, 50–4, 58, 71, 73, 74, 77, 82–3, 91, 93, 96, 99, 100, 110, 131, 133, 136, 147, 156, 159

technological optimism 47, 111, 149
technology/ies/ogical 4, 5, 15, 17, 19, 35, 39, 45, 47, 86, 90, 115, 136, 141, 144, 146, 147
Thunberg, Greta 2, 35
Thwaites Glacier 24
tipping points 24, 27, 32, 51
transformation/al 7, 8, 10, 39, 40, 43–9, 53, 54, 100, 131, 134, 151, 155, 156, 158, 164
trauma/s 8, 13, 27, 62, 99, 114, 122

vulnerable/ility 7, 19, 22, 26, 30, 50, 54, 92, 94, 95, 119, 120, 142, 143, 145

water (security) 13, 22–4, 28–30, 64, 96–9, 135
wellbeing 7, 10, 11, 23, 25, 65–7, 70, 83, 85, 98, 99, 109, 112, 118, 123, 124, 138, 140, 144, 150, 154, 155, 157, 160
withdrawal 9, 10, 35, 41, 48, 72, 89, 105, 111, 120, 122, 123, 125, 135, 141, 156

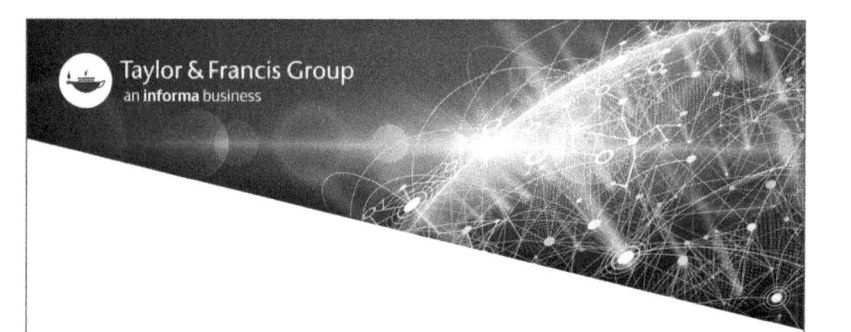